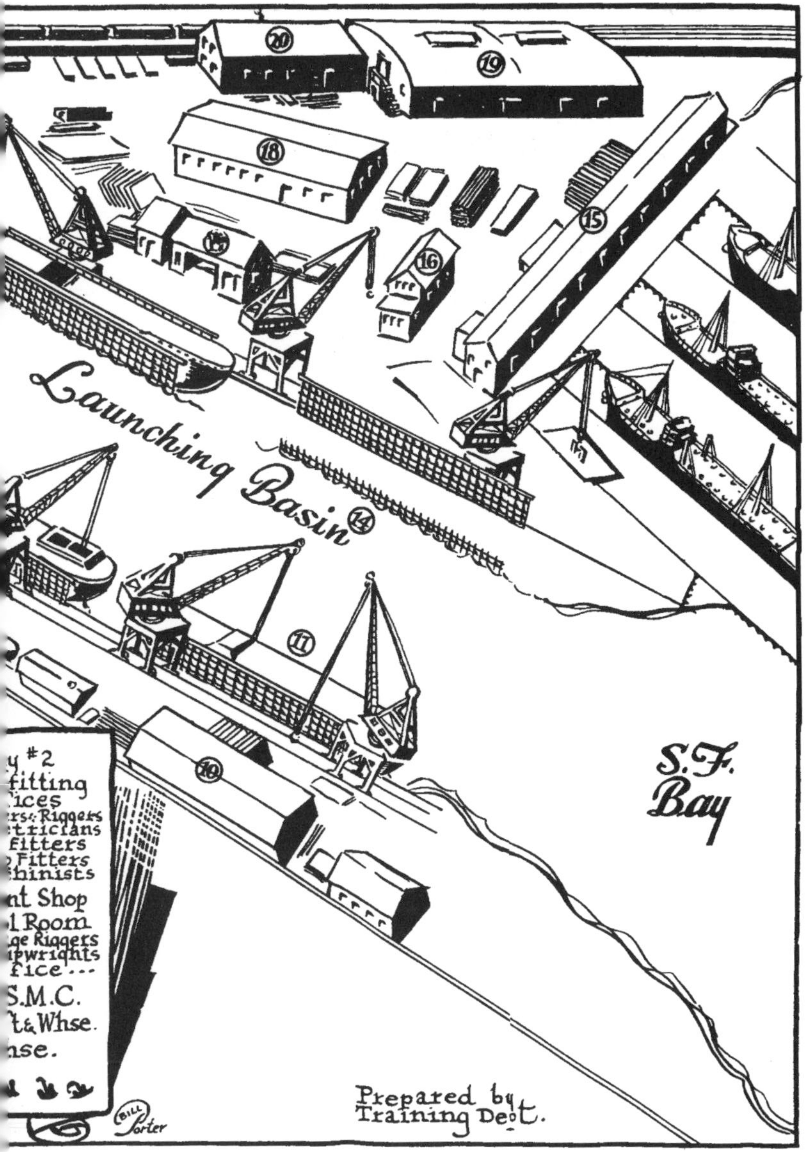

AF327142

20
19
18
15
16
Launching Basin 14
17
19
S.F. Bay
y #2
fitting
ices
rs & Riggers
tricians
Fitters
e Fitters
chinists
nt Shop
l Room
ge Riggers
pwrights
fice...
S.M.C.
t & Whse.
nse.
Bill Porter
Prepared by Training Dept.

STEEL SHIPS
AND
IRON PIPE

WESTERN PIPE AND STEEL COMPANY OF CALIFORNIA

THE COMPANY, THE YARD, THE SHIPS

PACIFIC MARITIME HISTORY SERIES

Number 5

Earlier books in the Pacific Maritime History Series are:

1. *Beyond the Lagoon*, by Lyndall Landauer.
2. *The Voyages Of the Ship Revere, 1849-1883* by M.R. Gleason.
3. *Clipper Ship Captain, Daniel McLaughlin and the* Glory Of the Seas by Michael J. Mjelde.
4. *The White Flyers,* Harvard *and* Yale, *American Coastwise Travel*, by George F. Gruner.

AMERICAN BUILDER

By the same author:

A Guide for Collecting L'Amour Paperbacks
The Butterfield Overland Mail
The America Of Eric Sloane, a Collector's Bibliography
Cruise Books of the United States Navy In World War II, a Bibliography

As editor:

With the Sommelsdijk *In the Far Pacific*

STEEL SHIPS AND IRON PIPE

WESTERN PIPE AND STEEL COMPANY OF CALIFORNIA

THE COMPANY, THE YARD, THE SHIPS

by

Dean L. Mawdsley

Associates of the National Maritime Museum Library
Pacific Maritime History Series
San Francisco
2002

Photo editor: Michael J. Mjelde

While every effort is made to obtain and reproduce the best quality photographs, there are occasions where the only photos available are those which are less than desirable, or are reproduced from a newspaper, or using old negatives. A number of these photos have nevertheless been included to provide a visual reference.

Copyright © 2002 by
Associates of the National Maritime Museum Library
Published by Associates of the National Maritime Museum Library
at The Glencannon Press

The Glencannon Press
P.O. Box 341, Palo Alto, CA 94302
Tel. 800-711-8985
www.glencannon.com

First Edition, first printing.

Library of Congress Cataloging-in-Publication Data

Mawdsley, Dean L.
    Steel Ships and iron pipe : the history of Western Pipe and Steel Company of California
/ by Dean L. Mawdsley.-- 1st ed.
        p. cm. -- (Pacific maritime history series ; no. 5)
    Includes bibliographical references and index.
    ISBN 1-889901-28-8 (alk. paper)
    1. Western Pipe and Steel Company--History. 2. Shipbuilding--California--History. I.
Title. II. Series

VM24.C2 M39 2002
338.7'6238204'09794--dc21
                                                    2002027661

Dedication

This history of a shipyard and the ships it produced
is dedicated to the many workers
who plied their trades in building these ships.

A special dedication is made to
the memory of Orrie Damewood
whose photographs sixty years later provide us with a
graphic picture of a shipyard at work.

Contents

ACKNOWLEDGMENTS

In 1985 I visited maritime author Kenneth Poolman in England and he was most helpful in securing information and photographs of the four escort aircraft carriers built at Western Pipe and Steel shipyard and then lend-leased to the Royal Navy. We have remained friends since.

A number of the workers including George Williams, superintendent of the outfitting docks, were interviewed, and all gave generously of their time and memories.

Research was carried out at the National Archives in Washington, D.C. and San Bruno, California, the Navy Department Library in Washington, D.C., the Nimitz Library at the Naval Academy Annapolis, Maryland, the South San Francisco Public Library, the San Francisco Maritime Museum Library, and the Imperial War Museum, London, England.

Of special help was Laura Brown, the librarian at the Steamship Historical Society's Library in Baltimore, Maryland. William Kooiman at the San Francisco Maritime Museum Library has been a source of information and also read this manuscript.

David Hull, Principal Librarian of the San Francisco Maritime Museum Library, provided both encouragement and suggestions for this project. The publications committee of the Associates of the Maritime Library; Peter Evans, John Kortum, Richard Geiger and Graydon Staring reviewed the final manuscript and provided further suggestions.

Appreciation goes to my wife, Mary Lou, who read the manuscript and gave many helpful suggestions.

Donald Reardon reviewed the manuscript and provided suggestions and photographic help from his perspective as a naval architect. Finally Walter Jaffee at the Glencannon Press worked diligently to produce this book.

The author extends his personal gratitude to all the above people for their part in the production of this project. He also realizes that many others, unnamed, played a part in the research extending over the past twenty years.

PREFACE

The author's interest in the Western Pipe and Steel shipyard dates back to the time of its greatest activity — the 1940-1944 World War II era. At that time, as a high school student, the author would ride in an automobile up the old Bayshore Highway to San Francisco. In South San Francisco south of Sierra Point the shipbuilding ways and outfitting docks were adjacent to the highway. There are recollections of the first C-3 type freighter, *Steel Artisan*, nearing completion. Then, for some reason, the superstructure was removed and a flight deck constructed. This was a source of much curiosity as the construction of escort aircraft carriers was still a military secret. The four aircraft carriers were completed and went off to war. Nothing more was seen or heard of these ships until the War was nearly over.

About thirty years later recollections of the shipyard and its activities were recalled as the author became more interested in the history of this region. The facts about the "mystery ships" were easily traced, and it was found that these ships were "lend-leased" to Great Britain and had illustrious careers in four oceans and the Mediterranean Sea.

From this beginning the author's interest widened to include the shipyard and all its progeny, including the World War I activity. During the last twenty years the author has accumulated information by research and interviews that has resulted in the writing of this book.

Abbreviations

AF of L	American Federation of Labor
AKA	Attack cargo ship
APA	Attack transport
CIO	Congress of Industrial Organizations
CVE	Escort Carrier
HMS	His Majesty's Ship
LMC	Labor Management Committee
LSD	Landing Ship Dock
NLRB	National Labor Relations Board
SHP	Shaft horsepower
SSF	South San Francisco
USMC	United States Maritime Commission
USS	United States Ship
WSA	War Shipping Administration
WPS	Western Pipe & Steel Company

PART I

WESTERN PIPE AND STEEL COMPANY AND ITS SHIPYARD

2 Steel Ships and Iron Pipe

CHAPTER 1
INTRODUCTION

South San Francisco for many years has had its slogan "The Industrial City" emblazoned on the south side of San Bruno Mountain in large bold letters. This tradition for industrial development goes back to the 1880s. In 1887 Gustavus Swift visited South San Francisco and saw the possibilities of developing meat packing facilities here. In December 1892 Swift with other investors opened a major meat packing facility which later became the Western Meat Company. Industrial growth followed with new plants for the Steiger's Terra Cotta and Pottery Works and W.P. Fuller

Company's paint factory in the next decade. However, growth was slow in the first few years.

During the 1890s and 1900s projects were underway to improve rail and water transportation to this region. The shorter direct mainline rail link was put through along the Bay's edge to San Francisco. The shallow waters along the shoreline were dredged to allow shipping into the new industrial area. The first movement of steel plants into South San Francisco came in 1903 when Pacific Jupiter Steel Company opened its steel casting plant. Other steel processing and fabricating plants soon began moving to the industrial area. Notable among these were Pacific Coast Steel Company in 1909 and Schaw Batcher Pipe Works in 1913. The Schaw Batcher site was later to be purchased by Western Pipe and Steel Company.

Western Pipe and Steel Company (WPS) was organized in 1907 in Los Angeles and the property of Thompson and Hoyle Company of Los Angeles, founded in 1886, was purchased. The main products at that time were metal casings for water wells and small caliber steel pipe. The Los Angeles business of WPS was expanded and the business moved to a new plant located at North Broadway and the Los Angeles River in 1908.

Construction of the launching ways began with the installation of pilings. Author's collection.

The first move of WPS into the San Francisco area occurred in 1910 when WPS purchased the Francis Smith Company of San Francisco, which was founded in 1854 in Grass Valley, California. This company made riveted pipes for the hydraulic mining of gold. At that time placer mining for gold was common, especially in the northern mines near Grass Valley. In the early 1860s the Francis Smith factory was established in San Francisco and it was this plant which WPS purchased. Shortly after the 1910 purchase, WPS purchased land across the Bay in Richmond, built a plant there, and moved the equipment of the old Francis Smith factory to Richmond.

It was in 1917 that WPS first moved into South San Francisco when they purchased the land and buildings of the Schaw Batcher Company, which was also in the pipe manufacturing business.

In that same year Schaw Batcher had obtained a contract from the United States Government to build twenty-two merchant ships. The first eighteen of these were of about 5,650 gross tons with a length of 410 feet. The last four were larger, being of 8,800 gross tons, but these four were cancelled at the end of World War I.

It took only three months to construct a shipyard on the South San Francisco site. The development of the shipyard construction can be

Riveting was standard practice in shipbuilding at the time, as can be seen by the riveting throughout this hull. Author's collection.

observed in a series of panoramic photographs taken and dated by San Francisco photographer, Gabriel Moulin. An October 1917 photo shows dredges at work and the beginning of shipways. By November there is less dredging activity and construction of the ways is more advanced. In January 1918 there is a keel laid and about March 1918 there are three ships under construction. A channel 7,000 feet long and 250 feet wide was dredged and four building ways were constructed on the edge of what was named Liberty Channel.

The keel of the first ship, the *Isanti*, was laid November 30, 1917. She was launched on June 2, 1918 and four other ships were under construction. Eighteen vessels of around 5,600 gross tons were launched between 1918 and 1920. The last of these ships, *West Carmona*, was delivered in October, 1920. These ships were of the standard riveted construction of the time. Side launching was used, as it was also in World War II at the shipyard. In the World War I launchings there was damage to the bilge keels of several ships during the side launching that necessitated repair in dry-dock. Later ships were launched without bilge keels which were added at the time of final dry-docking and fitting out.

The first eight ships were powered by General Electric steam turbines and the last ten ships by Joshua Hendy 3-cylinder triple expansion

Shipbuilding in 1918. Note keel and inboard scaffolding, lower left. Upper left, bulkhead, shell plating and double bottom are in place. Above, tank top plating is added to the double bottom. Author's collection.

Further along in the shipbuilding process. Above, plating on double bottom is almost complete. Right, a nearly completed hull in 1918. Author's collection.

The launching of the Isanti *on June 2, 1918. Left, a large crowd of guests and yard workers gathered for the launching,as final checks are made of the launching ways. Above, the* Isanti *just at the moment of launching. Both, Western Pipe & Steel Co., Author's collection.*

engines. The steam turbine engine frequently broke down, sometimes requiring that the ship be towed to port. These vessels were placed in the reserve fleet in the early 1920s and were scrapped by 1930. The ships powered with triple expansion engines were active into the 1940s and took part in World War II.

It is unclear at present whether WPS management played any part in the World War I shipbuilding. It would appear that Schaw Batcher obtained the contract from the United States Government and that this contract together with the business, the land and buildings was purchased by WPS. This is substantiated by the fact that vessels are numbered 1-18 in the WPS hull number sequence.

During World War I, WPS continued operation at its Richmond plant, making pipe and metal tanks for the United States Government. In 1921 the Richmond plant of WPS was dismantled and moved to the South San Francisco site that had been purchased in 1917.

In the mid-1920s the outfitting dock and launching basin at South San Francisco were used to tie up inactive Standard Oil Company tankers.

The determination to defeat Germany in World War I is evident in this hull plating about to be lifted aboard a ship under construction. Author's collection.

Laidup Standard Oil tankers at the WPS South San Francisco facility in the 1920s. San Mateo County Historical Association.

This apparently did not involve closure or preservation (mothballing) measures.

During the 1910-1920 decade WPS had been spreading its operations. In 1910, a factory was established in Taft, California to serve the oil industry by producing steel pipes and containers. In 1913 a plant was built in Fresno. Later, in 1942, a new and enlarged factory was built at that city. In 1915 a factory was established in Phoenix to manufacture pipe for the agriculture and oil industries of Arizona.

In the 1930s, WPS established temporary manufacturing facilities at Grand Coulee, Washington and Shasta Dam in California. These plants manufactured large penstocks for the hydroelectric generators. These penstocks were large caliber pipes that carried the water from the damned lake to the generating turbines and were too large to transport, necessitating their manufacture on site. In the 1930s, WPS was making large caliber pipes for the water conduit from Hetch Hetchy Dam to the Crystal Springs Lake on the San Francisco Peninsula with its subsequent extension from San Mateo to San Francisco. Also in the late 1930s a fabrication plant was setup in Seattle, Washington for work on a pipeline in Everett, Washington.

In World War II, WPS also operated a Shipbuilding Division in San Pedro, California. The San Pedro Shipyard constructed ships for the U.S. Navy and Coast Guard. The ships built at this yard were icebreakers, destroyer escorts and landing ships. The details of this operation are beyond the scope of this study.

Seventy years after its beginning, in 1983, the site of the old Western Pipe and Steel Company plant was sold and all the buildings were razed. Now, only a few remnants remain: the launching channel and rotting dock pilings. These are the only reminders that once this facility was bustling with activity and employed several thousand people. Many of the officials and workers at Western Pipe and Steel have passed on, and many others are getting to advanced years. Soon all recollection of the Yard and its contribution will be gone.

In the redevelopment plans for this land are dreams of large hotels, office buildings and a convention center. Never again will the forms of large ships rise from the building ways or the workmen with their hard hats stream in and out with the changing of work shifts. Times have changed in South San Francisco and no one can turn the hands of the clock back; however, we should not forget the contributions made here during the bustling wartime periods of 1916 to 1920 and 1939 to 1946.

Chapter 2

Administration

The corporate headquarters of WPS was located at 200 Bush Street in San Francisco. This office provided general direction for the WPS plants at Fresno, Mayfield, San Pedro and Phoenix as well as the South San Francisco operation. Contracts were negotiated by the officers and financial control rested at this headquarters and an annual report for stock holders was prepared here. The medallion insignia used by WPS had as its centerpiece the Mechanics Statue located on Market Street near the corporate headquarters.

H.C. Tallerday was president of WPS until 1943, and then became chairman of the board when L.N. Slater succeeded him as president. Vice presidents were R.D. Plageman, L.W. Delhi, Francis S. Howard and William H. Schutte. Reese Tucker was the secretary and treasurer. The board of directors was made up of W.G. Aldenhagen, J.J. Baumgartner, A.M. Duperu, O.B. Perry and John B. Beman in addition to Tallerday, Slater, Howard and Plagemen.

In the 1930s WPS maintained a small but efficient employment office under its employment manager, Doc Hart, who had been working at WPS since the 1920s. Doc Hart's nickname stemmed from the fact that he was a pharmacist, and in addition to his duties in the employment office he also ran the first aid facility at the South San Francisco plant. In early 1941 Tom Nicolopulos was added as clerk and assistant employment manager. Approximately 2,600 people were employed at WPS in 1941.

As the work force rapidly expanded in late 1941 the duties of the employment office became more complex. Its name was changed to the Personnel Office. Doc Hart remained as personnel director, but was later replaced in August 1943 by William Farmer. Tom Nicolopulos remained as assistant personnel director.

William Farmer at his desk. WPS News *courtesy of author.*

Sam Crow and camera caught in a rare moment on the other side of a lens. WPS News *courtesy of author.*

Shortly after the Pearl Harbor attack, the government imposed increasing security measures to prevent spying and sabotage. One of the first measures instituted was the identification and investigation of all workers. An identification division was set up in the personnel department under Sam Crow, the plant photographer. Initially he photographed all men then working and subsequently photographed all new employees. It was necessary to make two prints of each employee, one for the identification badge and one for the files. Crow made over 102,000 prints in the war years. Later the FBI also required fingerprinting, with a copy retained in the plant and a copy sent to the FBI in Washington DC. As plant photographer, Sam Crow took photos of accidents, bond rallies, special events and publicity photos. Also many photos were done for the plant newspaper, *The Western Pipe & Steel News*.

There was never a proven act of sabotage at the shipyard. However, there were two suspicious fires in the yard and Orrie Damewood recounted having to go out to photograph the site on these occasions. One instance involved a fire on a ship-way underneath a hull which was under construction. The only damage in this case was some buckling of ship plates requiring replacement.

With the increasing shortage of many items wartime rationing was started. Workers in defense plants were allowed supplemental rations of gasoline, tires and shoes as needed. A special rationing office in the personnel department was set up to procure and monitor these supplemental rations.

Selective service deferments were granted to essential workers. The personnel department had a division to monitor deferrals. Every six months Tom Nicolopulos went to Sacramento to review the list of deferments with the state Selective Service office and secure renewal of them.

Hiring and termination procedures were, of course, part of the personnel department function. Late in the war the government froze workers in their jobs and special releases had to be obtained before a worker could leave and work elsewhere.

On occasions workers reported grievances to their union. Tom Nicolopulos had the duty to investigate and mediate such complaints. Union cooperation in such cases was excellent and no work slow-downs or strikes resulted.

The plant safety office headed by Jim Kelleher was another major division of the personnel office. This office had functions of prevention,

TIRE RATIONING OFFICIAL TELLS PROCEDURE

Workers Must Certify Needs to Labor-Management Committee

By KENNETH R. LOWELL
(State Tire Rationing Board)

Because of the grave rubber shortage confronting our country today, the tire rationing regulations are necessarily strict.

In order to supply recaps and grade II tires to defense workers as adequately and fairly as possible, it is necessary in each case that the automobile utilized for transportation from the worker's home to his place of work in a passenger automobile be essential.

Concern for safety was critical. Every issue of the company paper featured articles, cartoons such as above and photos such as the demonstration of the effectiveness of safety glasses, below. WPS News *courtesy of author.*

investigation and reporting of accidents. Articles were regularly run in the *WPS News* to promote safety for workers. Many of these articles were authored by the plant safety engineer Jim Kelleher himself. When accidents did occur, reports were obtained from the injured party and his supervisor by the safety office and forwarded to the yard superintendent. Also a safety store was established in the cafeteria to sell safety hats, shields and shoes as well as books on related subjects.

Staffing was tight and to meet the shipbuilding schedules many more workers were required and had to be actively recruited. WPS did not mount a nationwide recruiting program comparable to that of the Kaiser and Bechtel Corporations. The WPS personnel needs were not as massive as were those of the other larger yards. WPS emphasized contacts in the San Francisco Bay area, especially on the Peninsula from San Francisco to San Jose. Many leads were obtained from workers in the plant. During the war period many old South San Francisco families were represented in the yard work force, and it was not unusual to have several members of a family working at WPS.

An August 1941 report showed the following distribution of residences of the 2,655 WPS employees: San Francisco, Daly City and

Colma 56.3%, South San Francisco 14.6%, Mid-peninsula 14.8%, East Bay 7.8%, Santa Clara County 2.8% and Brisbane 2.6%.

As the personnel needs became more pressing Ted Smyth was appointed as full-time recruiting coordinator. An example of the recruiting is seen in a report of recruiter Marcial Barellano. A Filipino himself, Barellano contacted Filipino groups in San Jose, Salinas and Stockton. He discovered that the Filipinos were making more money in their present agricultural occupations, so very few prospective workers were found. Several local men such as Mr. Hardy, Mr. Holston and Mr. Cargill were hired to make contacts and bring in workers. They made the rounds of employment offices and even approached people on the street. Contacts were even made with prisoners about to be released from the San Francisco County Jail in San Bruno. Some of these men were brought in to work for WPS. At times Allied merchant seamen, mostly British, Greek and American, came to the yard to work during leave time. U.S. Navy personnel from the Tanforan Navy Base also came to work for WPS during leave periods. The union cooperated by extending free work permits for these service people. The union also did some recruiting in the skilled trades.

EMPLOYMENT

By SAM CROW, MGR.
Identification Dept.

Badges are made and issued to each employee to be worn on the "left side of the chest" at all times. If you find your paycheck is held up, you can get same by going to the pay office and presenting your badge. The slogan from now on is "No Badge, No Paycheck."

As the nation became taxed by the great manpower needs of the military and the expanding wartime industries an increasing number of women and minority workers were employed at WPS. The author has seen no statistical breakdown of numbers, but it is abundantly clear from the *WPS News* that this was occurring. Women were working in almost all of the crafts, but were most numerous as welders and tackers.

Another outstanding example of recruiting was reported from Oakland where a black minister succeeded in bringing in a number of members of his congregation to work for WPS.

Chapter 3

The Shipyard In World War II

With the passage of the Merchant Marine Act of 1936 Congress reflected the national desire to reestablish a shipbuilding industry in the United States while affirming America's position as a leading trader in international ocean-going commerce. The objective was to have at least 50% of U.S. cargo carried by U.S. ships. During World War I there was a wartime flurry of shipbuilding; however, following the war there was a glut of ships, and the United States was unable to compete in building and operating its ships. By the mid 1930s the United States Merchant Marine was in a depressed state as was the general economy. With the United States Maritime Commission (USMC) providing subsidies for ship construction as well as ship operation there was help for both the general economy, with the creation of more jobs, and for the merchant marine thus making it more competitive with foreign shipping.

During the mid 1930s, under the Roosevelt Administration, H.G. Tallerday, president of Western Pipe & Steel Company, was appointed to the National Labor Relations Board. In this capacity he traveled to Washington, D.C. by train as often as once a month and acquired many friends, among whom was Admiral Emory Land, chief of the USMC. Tallerday had worked in ship construction during World War I in San Pedro, California and wished to return to shipbuilding. The contacts in Washington, D.C. undoubtedly assisted Tallerday and WPS when submitting bids for shipbuilding. In Mr. Tallerday's mind also was the fact that the WPS's site in South San Francisco had been the site for shipbuilding in the World War I era. Thus WPS had a plant with existing shops, rail transport connections, and a waterside location suitable for embarking on a shipbuilding program.

Prior to commencing reconstruction of the yard, WPS had a contract with the Navy to build eleven 500-cubic-yard dump scows. These were

Dump scow number 7, one of several 500 cubic yard dump scows constructed during the late 1930s. Author's collection.

A U.S. Navy light cruiser tows one of the dump scows out the Golden Gate. Western Pipe & Steel, author's collection.

built for service at Midway Island. A company photo shows one of these scows under tow by a light cruiser passing through the Golden Gate.

During the first two years of its existence (1936-1938) the USMC developed standard designs for three types of dry-cargo ships, the C-1, C-2, and C-3. These ships were designed to be competitive in speed and cargo handling with foreign-built ships. Their crew accommodations, safety features and general construction were to be superior and more expensive, but this would be offset by Federal subsidies for construction and operation. These superior ships were designed also with a wartime emergency in mind so they could readily be converted to fleet auxiliaries for the United States Navy.

In early 1939 WPS was among several firms submitting bids for construction of the C-1 type cargo ship. The awarding of the USMC contract was delayed because the WPS bid was one of sixteen bids and was higher than that of a Southern California firm. After further consideration WPS was awarded a contract in October 1939 to build five C-1 cargo ships at a cost of $10,635,000. At the same time, they received $400,000 for construction of the shipyard itself. This amount would cover dredging a channel into the outfitting and launching basin, construction of

WPS shipyard September, 1940. C-1 freighters under construction on two ways north of the launching channel. The first ship, American Manufacturer *is at the outfitting dock. The* American Leader *and the* American Builder *are shown on the ways. Author's collection.*

View of newly constructed ways on south side of launch basin, September 24, 1941. The hull of the Steel Artisan *is shown on way number 1 on the far left. The* American Packer *is at the outfitting dock. The keel of USS* Cascade *is seen on way number 3 to right of launching basin. The gantry crane in the foreground is being assembled and pile drivers are working in the basin to the left. Author's collection.*

two shipbuilding ways on the north side of the launching basin, and construction of nine buildings including offices, warehouses, shops and a restaurant. Yard construction was to start promptly so that the first keel could be laid in three months

Chapter 4
The Management

H.G. Tallerday. WPS News *courtesy of author.*

Howard G. Tallerday: HGT came to work for the Talbot brothers, owners and founders of WPS about 1906, soon after the company's founding. The Talbots retired from WPS in the early 1920s leaving HGT in charge. At the time of the rebuilding of the shipyard just prior to World War II, HGT was president of WPS. In November 1942 he became ill and did not reappear at ship launchings or programs until May 1943. At that time he was chairman of the board and L.N. Slater was president. HGT remained as chairman of the board until his death January 26, 1946. Soon after, WPS was sold to Consolidated Steel Corporation and the new firm of Consolidated Western Steel Corporation emerged.

L.N. Slater: During the period of building of the shipyard up to early 1943, LNS was a vice president. He was elevated to the presidency in early 1943 and remained president until sale of WPS in January 1946. Slater had been with WPS since 1913 starting as a salesman in the Sacramento Valley. In 1918 he was made manager at Bakersfield and Taft. Transferred to Los Angeles, he was assistant manager of the Los Angeles office in 1924 and manager in 1926. He became a member of WPS board of directors in 1926.

L.N. Slater. WPS News, *courtesy of author.*

O.B. Kibele: OBK had a background as a machinist and engineer. He had an inventive mind and held a number of patents. In World War I, OBK was superintendent of the Schaw Batcher shipyard on this same location. In the years between wars he owned a boat building business in San Pedro and was a Harbor Commissioner of Los Angeles Harbor. When the WPS yard was rebuilt he was brought back as superintendent. He was an excellent organizer and judge of people and delegated authority well. He

O.B. Kiberle. Courtesy of author.

T.R. (Ted) Rooney. Courtesy of author.

Harvey   Clarke, WPS News, *courtesy of author.*

was approachable by anyone in the yard and was willing to make decisions. OBK had a kind fatherly attitude to his employees, but at the same time could be tough when that was required. OBK died suddenly on October 7, 1944, at age 76 and was followed as general superintendent of the shipyard by T .R. Rooney.

T.R. (Ted) Rooney: TRR was shop superintendent during most of the World War II shipbuilding. In February 1943 he became superintendent of the building ways as well as the shops. TRR was elevated to the position of general superintendent of the shipyard in November 1944 following the death of O.B. Kiberle. He was an able coordinator and much liked by men working for him.

Harvey Clarke: An early employee of WPS, Clarke had worked at least thirty years for WPS by 1942, at which time he was assistant shop superintendent under T.R. Rooney. He later became shop superintendent when TRR was elevated to general superintendent of the entire shipyard. HC was a demanding and critical boss, often antagonizing workers under him. His life was ended by suicide after the World War II period.

George Williams: GW started to work for WPS in 1926 as a machinist. He was soon promoted to machine shop foreman. In 1928 he headed the on-site WPS plant at the Grand Coulee Dam for installation of the penstocks. On his return to the SSF plant the shipbuilding program was getting underway and GW was appointed as superintendent of the outfitting docks. This position was expanded to include outfitting work done both at the SSF plant and on the San Francisco waterfront at piers 27, 88, 90 and elsewhere. GW was known to be a strict disciplinarian and did not tolerate slackness on the job or inferior work. He became assistant to the Vice President of United States Steel in 1951 and was responsible for production at the SSF, Berkeley and Fresno plants, all of which had been WPS plants prior to the 1946 sale. GW retired in the 1950s.

Cecil Patterson: CP started work at WPS in 1930 as a maintenance machinist. He spent eight years in the shop working on the machinery. In June 1938 he went to the Grand Coulee plant where he stayed until December 1940. Returning to South San Francisco in January 1941, he worked on the construction of the shipyard. Among his tasks was the setting up of the three traveling cranes along the two ship-ways on the north side of the launching channel. CP worked as yard maintenance

George Williams. Courtesy of Richard Williams.

mechanic until he was called upon to take charge and coordinate the completion of the outfitting of USS *Chandeleur*. Following completion of this ship, CP remained in the outfitting department as assistant to superintendent George Williams. CP continued in the shop after the war and retired in 1972.

Foremen: On each ship there was a coordinator to check on work progress, coordinate materials necessary for the work at hand and report to the outfitting superintendent. Each craft had a foreman in charge of the men in his craft on all the ships then outfitting. On each of the hulls during outfitting, the trade on that ship was under the charge of a leaderman. The trades represented on the outfitting hull included machinists, shipfitters, pipe fitters, steam fitters, painters, sheetmetal workers and electricians. Among the foremen were Henry Hook, electrical foreman; Ed Warne, machinist foreman; Jack Warner, pipefitting foreman; Norm Taylor, sheetmetal foreman; Ed Brewer, shipfitter foreman; and Louis Reiche, painting foreman.

Mold Loft: The mold loft is one of the key units in the function of a shipyard. In the WPS yard the loft was located in a warehouse building north of the outfitting dock near the Southern Pacific railroad line. In the early months of shipbuilding, Ray McCord was in charge of the mold loft and later his job was taken over by Clyde Vassar. In the mold loft the blueprints of the ships were translated into full size patterns or templates for each part. This process required a large open area where patterns could be laid out. Tracings were then made on thin wooden sheets and cut. The patterns were then taken to the shops where they were used to make up the individual ship's parts such as plates, frames and structural supports. Frequently the pattern would break and the mold loft was always ready to produce a new template for any part. About thirty-five people were employed in this department.

Phillip Stroupe and Orrie Damewood: Phillip Stroupe was a commercial photographer in San Francisco who had a contract with USMC to provide monthly progress photos of ship construction. These photographs were to be 8 x 10 inch photographs, one bow view and one stern view, each month of every ship under construction.

Orrie Damewood came to work for WPS in September 1941. He worked as an expeditor of equipment. Orrie became acquainted with

Phillip Stroupe, who did not like the heights he would have to climb to take photographs. He described being lifted in a bucket to a considerable height by a crane to get better angles for his photographs. OD began taking these monthly progress photographs and soon was taking all the photographs and sending them to Phillip Stroupe for processing. Damewood worked at WPS until a few weeks after the war ended in September 1945 when he went into partnership with Stroupe. During the war years over twelve hundred progress photographs were taken. These photos remain as a great storehouse for historians. Stroupe died shortly after the war, but Damewood continued as a professional photographer in San Francisco until his death in 1989.

CHAPTER 5

LABOR AND MANAGEMENT

During the construction of the yard trouble developed on the labor front. Unions were pressing to organize the previously nonunion WPS plant, and an organizing election of plant workers was proposed by the American Federation of Labor (A.F. of L.) Metal Trades Council. This election was held in late 1939 and the workers chose them as their union. The picture was further clouded by a suit by the CIO with the NLRB challenging the A.F. of L.'s right to represent WPS workers. These uncertainties caused a temporary suspension of work on the shipyard facilities. This delay in October 1939, however, proved short and within a month yard construction resumed with the A.F. of L. Metal Trades Council representing WPS workers as well as workers of most other Bay Area shipyards.

The government, the shipbuilding industry, and the unions had a master agreement for each of the four major areas, *i.e.* the Pacific Coast, the Atlantic Coast, the Gulf Coast and the Great Lakes region. During the remainder of the prewar and World War II period there were no further labor delays or strikes. In 1946, after the war had ended, the first labor strife arose with a strike involving the machinists of Lodge 68.

The unions under local executive Ed Rainbow helped recruit skilled workers, mediated worker-employer disputes and promoted employee benefits. They worked as did everyone else with the winning of the war as their primary goal.

The establishing of Labor-Management Committees (LMC) in defense plants was pressed by the government to increase production and morale in industry. By September 1942 over 1,600 had been established throughout the country. This committee was already functioning at WPS in September 1942 and had representatives from the workers, the union, management and the USMC. Initially the LMC was under the direction of William Farmer, a plant official.

Union official Ed Rainbow calling for an end to "loafing and absenteeism" at the shipyard in February, 1943. WPS News, *courtesy of author.*

Left to right: J.H. Gallivan, F.J. Swaney (who made the presentations), Ed Twig, Rico Ferrario, R.E. Lehtinen and Lloyd Fry receive War Bonds for their labor saving ideas. WPS News, *author's collection.*

A Suggestions subcommittee was set up with Irving Larsen as chairman. Workers were to submit labor saving ideas to be judged by the committee. Four awards were made monthly with prizes of $25 to $100 in war bonds or cash. The first of these suggestion awards was made in September 1942. The manual used throughout the yard for construction of the C-3 type of vessel was the result of one of these awarded ideas. The USMC also printed this manual for distribution to other shipyards. In February 1944 the LMC was chaired by of A.W. Tate of the engineering department.

Many problems came under the purview of the LMC. Transportation of the workers to and from work always presented a problem. Car pooling arrangements were publicized and given assistence. A special company bus service from San Francisco, Daly City and Colma was begun and operated at cost. This service was terminated in May 1945 when use was down and financial losses were increasing. Special arrangements were made with the Southern Pacific Railroad for trains to stop at Butler Road near the Yard to facilitate worker transportation.

Food was always a source of complaint and many complaints in this area were heard. These were negotiated where possible with the cafeteria management. Plans were submitted for a new enlarged cafeteria and for mobile catering units for distant areas such as the outfitting docks. Eating schedules and crowding at noon were always a problem. Many complaints were heard about the inferior cigarettes carried in the cafeteria. But little could be done about this because the better cigarettes all went to the military services.

The LMC coordinated war bond drives and appointed special subcommittees to promote bond sales and special events for each drive. On occasions a large companywide dance was held in San Francisco's Civic Auditorium to help both morale and the war bond drives.

Western Pipe & Steel's third Victory Bond Dance was held at the Civic Auditorium in San Francisco. Queen Clair and her maids of honor enter on the arms of company officials. WPS News, author's collection.

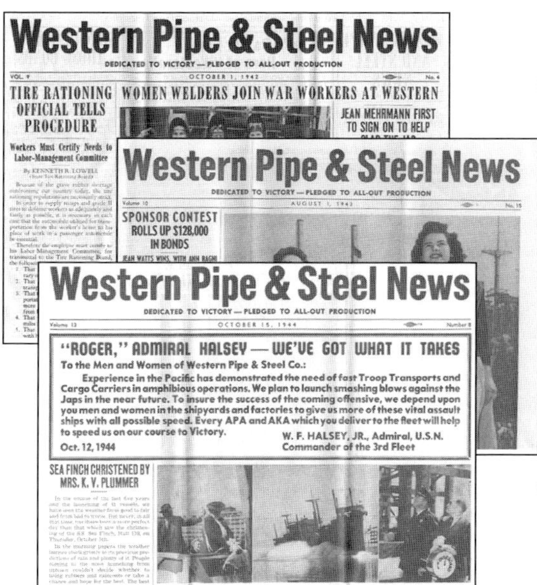

Western Pipe & Steel News *was published bi-weekly and covered all the activities of the shipyard. Author's collection.*

CHAPTER 6

WESTERN PIPE & STEEL NEWS

Prior to the shipbuilding era, WPS had a company bulletin published quarterly. With USMC encouragement, WPS started publication of the *WPS News* twice a month to improve plant moral by reporting news and spreading a great variety of information. The *WPS News* had offices in the yard, but was under direct control of the San Francisco corporate headquarters. The editor was Dick Prosser and associate editor was Joe Tessier. The *News* commenced September 1, 1942, and the last issue the author has seen is dated November 1, 1945. By reading through the *WPS News* one can still get a feel for the yard's activity. Early issues portrayed an expanding activity with much discussion of training new workers. In contrast during the last six months a definite feeling is obtained of the slowing pace. Many employees were leaving voluntarily or being laid off.

News of yard activities was featured and a prominent coverage was always done on ship launchings. This included photos and a feature on the ship's sponsor. A special feature issue occurred when USMC chief

Launchings were featured in detail with photo collages showing key events. WPS News, *courtesy of author.*

TRAINING CLASSES NOW UNDER WAY

The training program for Shipfitter and Flanger trainees started at Western Pipe & Steel on September 28. Classes for Swing Shift are from 2:30 to 4:30 p.m. and for Day Shift from 4:30 to 6:30 p.m.

Each trainee attends either a Monday and Wednesday class or one on Tuesday and Thursday. Any trainee who would like to study more is welcome to take two subjects totaling eight hours per week.

For men who want to learn blue-print reading faster there are classes running five days a week.

At present, classes are being held in Working Lines and Blue-Print Reading and in Shipfitting and Flanging Practice.

Future classes are planned in the above subjects and in Advanced Individual Blue-Print Reading, and in Mold Loft Practice. Sign-up for these four subjects is now under way in the Berthing Office (located between Ways 3 and 4). Enrollment is open to all, and any man who is interested is urged to get his name in at once.

Training was continual, with a column devoted to it in almost every issue. WPS News, *author's collection.*

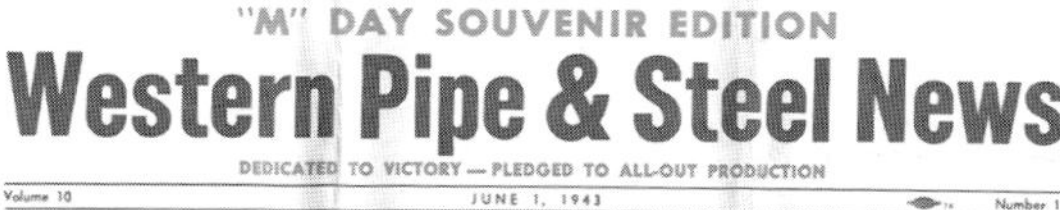

"M" DAY SOUVENIR EDITION

Western Pipe & Steel News

DEDICATED TO VICTORY — PLEDGED TO ALL-OUT PRODUCTION

Volume 10   JUNE 1, 1943   Number 11

ADMIRAL LAND PRAISES W. P. & S. WORKERS AT PRESENTATION CEREMONY

"MIRACLE OF SHIPBUILDING" SAYS CONSTRUCTION DIRECTOR FLESHER

SPEAKERS SAY AWARDS WELL MERITED

Sunday, May 23rd, goes down in our book as the greatest day in the history of Western Pipe & Steel Company, and to many who were present at the presentation ceremonies and launching it was the outstanding event of their lives.

We met Admiral Land, and we liked him. Here is no brass hat or stuffed shirt, but a genuine human being of the earth, earthy, and a fellow worker whose every thought and effort is devoted, just as our own best ability is devoted, to the great task of winning this war.

We heard Carl Flesher, who is responsible for all Maritime Commission construction on the Pacific Coast, praise our achievements, and we liked him and were pleased at his acknowledgment of the worth of our endeavors.

We rejoiced at the appearance of "Uncle Howard" Tallentday at the microphone, and relished with deep satisfaction the compliments he voiced and the thanks which he generously declared were due to every employee.

We were proud, too, when President L. N. Slater, in characteristically easy manner, addressed us as "folks" and speaking for both management and workers, paid tribute to Mr. Slater and his entire staff.

A visit and awards by Admiral Land warranted a special edition. WPS News, author's collection.

Admiral E.S. Land visited the plant, witnessed a launching, and presented WPS with the USMC "M" award. Feature articles were done on personalities such as Sam Crow, the plant photographer. Other articles dealt with shipbuilding activities and WPS ships already in service. Safety articles were a regular feature and some were accompanied by photos or cartoons.

Many departments had their own regular gossip column reported by a member of that department. Such columns came from the outfitting docks, shop, engineers, employment department, purchasing department, payroll department and many others. As the roll of woman workers grew, a special column for women appeared. This was titled WOW (Women of WPS) and was written by Meryl Shroyer. This column contained special safety suggestions as well as news of women's activities.

In the *WPS News* could be found details of classes available for workers to help them improve their skills. In the September 1942 issue there appeared an announcement of a class for shipfitter and flanger trainees. Classes in blueprint reading and mold loft practices were noted. In November 1942, reflecting the growing number of women employed in the yard, a course was offered in basic shipbuilding nomenclature and practice. Activities of the LMC were carefully covered.

WOW!

(WOMEN OF "WESTERN")

By MERLE J. SHROYER

It's an interesting fact that in proportion to the number of women employed in the various crafts in the Shop and Yard there are well over two hundred whose husbands are in the service of Uncle Sam. The general attitude of the girls seems to be to do their part in the building of ships as a means toward a quick victory and the home-coming of their soldier and sailor husbands.

* * *

Wanderings around the Yard: There's a new helper wearing good looking open toe sandals—she's pretty too, with her curls unconfined blowing in the breeze! So—she forgot her bandana and she's going to get safety shoes right after work! . . . Have you had occasion to notice the gleam in the eye and the blush on the cheek of ALICE CHAVEZ these days? Uh huh—a new romance . . . Those girls in the dock tool rooms, HESTER McCOWN and EVELYN WELLHAUSEN, are busy but never fail to display a friendly smile at each customer . . . I've missed seeing LILLIAN MURPHY, that little hard working pipefitter helper and I learn that she's very ill. Let's wish her a speedy recovery and hope that she'll

Women of Western Pipe and Steel was a column devoted to women. WPS News, author's collection.

The WPS basketball team was featured in one issue for winning the league championship. WPS News, *author's collection.*

Publicity was given for a variety of events such as dances and other social events. Many articles appeared on featured drives such as the sale of war bonds, recruiting of blood donors and the collection of books for ship's libraries.

Features were done on sporting teams and their results. The bowling, baseball and basketball teams were always active in their respective seasons. In all, the *WPS News* promulgated a great deal of information and promoted better morale for the entire plant.

Judging by the lack of smiles, one might guess the team lost on this night. In any case, the bowling team was listed in most issues. WPS News, *author's collection.*

Admiral Land presents the "M" pennant and the Victory Fleet Flag to the shipyard. Left to right: L.N. Slater, Admiral Land, Howard G. Tallerday and L.W. Delhi. WPS News, *author's collection.*

CHAPTER 7
AWARDS

Awards were used by the government to stimulate production and to promote war bond sales. There is no question that these awards were coveted and did promote their objectives. In 1942 the *WPS News* was advocating increased production to get the USMC "M" flag for the yard. The yard had to be ahead of the contract schedule to be eligible for this "M" award. Finally the day came on May 23, 1943 when Admiral Emory S. Land himself visited the yard and presented the USMC "M" flag. This rally, which also included launching the *Sea Devil*, was a highlight in the yard's history and was attended by a large crowd. In

45

Shipyard rally for the 7th War Bond Drive. This was in conjunction with a ship launching and presentation of a gold star for the yard's "M" flag. Photo by Orrie Damewood, author's collection.

Workers received merit badges for productivity. WPS News, *author's collection.*

addition to the flag for the plant, individual workers received a merit badge. To prevent a work letdown after the award, a review of production was done each six months and an extra star was added to the "M" flag for subsequent awards. WPS received additional awards in December 1943, June 1944, and December 1944. The June 1944 award was presented personally by Admiral Vickery, Admiral Land's second in command, during an inspection of the yard. By mid-1944 the "M" flag had four gold stars on it.

Admiral Vickery, second in command of the Maritime Commission at an awards ceremony at Western Pipe & Steel. WPS News, *author's collection.*

War bond sales were promoted by the Treasury Department with the award of a Minute-Man Flag. The first bond sales drives at WPS lagged and no Minute-Man Flag was awarded until August 1944. To win this flag it was necessary to have ninety percent of the personnel signed up for payroll deduction purchases of bonds and to pass the assigned quota for each bond drive. Bond drives later in the war sold better and the Minute-man flag was awarded after each drive.

In October 1945 WPS received a special safety award from the Department of Labor for reducing accidents forty percent in the first six months of 1945 compared to the same period a year earlier.

All these awards had their intended results in boosting morale and promoting a sense of cohesiveness for the entire operation.

The Minute Man flag was presented by Edward Massa of the Treasury Department, left, with Rollie Cate holding up the other end. WPS News, *author's collection.*

CHAPTER 8
SHIP CONSTRUCTION

The intricate process of building ships began many months before the first keel laying with the decision in the WPS corporate offices to go into shipbuilding. Corporate president H.G. Tallerday then made contact with USMC offices in Washington, D.C. to start the process of submitting a bid for ships the USMC desired built. In the case of WPS these were C-1 and C-3 type standard USMC freighters. The plans and specifications obtained were studied by the corporate officials with the engineering department to make cost estimates required before a formal bid could be submitted to the USMC.

A delay occurred at this stage while the USMC considered many factors before awarding the contract. The foremost consideration was cost so it had to compare the bids of several shipbuilders. In 1939 the USMC studied the local unemployment situation hoping to increase employment and lessen the effects of the depression from which the nation had suffered during the 1930s. The ability of the company to deliver a sound well-constructed ship was especially important since WPS was not yet in the shipbuilding business. Although WPS lacked experience, it had an advantage. The planners in USMC realized that a national emergency was in the offing and wished to establish new shipyards to increase the nation's shipbuilding capacity. For this reason the USMC was willing to make money available to construct these new yards. Finally, in August 1939, the WPS bid was accepted, and more definitive plans could be started.

Details went to L.W. Delhi, vice president in charge of shipbuilding. The engineering department under Thomas McMullen had then to take the plans and work up detailed drawings and specifications. This information was used by the purchasing department under Frank Summers to order promptly the many materials and parts necessary so that materials would be at hand and there would be no lag in construction. The plans were also sent to the mold loft under the jurisdiction of Ray McCord, where patterns were made for thousands of ship's parts. These patterns were sent to shops headed by Ted Rooney where fabrication and assembly into transportable units was carried out. The transportation department with the aid of cranes and flatbed vehicles took parts from the shops to the assembly area adjacent to the ship-ways.

In building an individual ship, even though much preliminary work is needed, the laying of the keel is considered to be the birth of the ship. Shipwrights under the leadership of Jack Wilson prepared the blocks and supports on the launching way so that the large weight of the completed hull could be supported and launched. Next, the keel plates running the length of the ship's center bottom are laid in place, aligned and welded together. The double bottom, previously made in sections in the shop, is placed and welded to the keel. From this bottom arise the transverse bulkheads supporting steel beams and plates. At this stage many skilled craftsmen are required. Shipwrights control the placement of all sections. Plate hangers put the plates in place. Shipfitters make final adjustments and do detail work. Tackers and welders weld the parts together.

Laying keel for SS Sea Pike *on March 5, 1942. Note scaffolding inboard side of ship and previously launched ship at outfitting dock in background. Western Pipe & Steel photo No. 396, author's collection.*

Left, keel and double-bottom construction on SS Steel Artisan, *April 1941. Right, Hull construction on SS* American Packer *in March 1941 showing bulkheads and tween deck plating. Western Pipe & Steel, left photo No. 225, right No. 213, author's collection.*

Pipefitters and machinists do the detail work on final lines and engine supports. Electricians are needed for the many electrical lines running throughout the ship. As each section is completed the WPS inspection department makes a careful examination prior to the inspection by the USMC.

For centuries the launching of a ship from dry land into the sea in which she will spend her useful life has been both a symbolic act and a practical test. It is then for the first time that the new hull has its first test of seaworthiness, *i.e.* the ability to float and not leak. This ceremonial act was performed forty-eight times at the WPS yard between 1940 and 1945.

Probably the most unusual aspect of launching at WPS was the side-launching. Because of the limited land and waterfront, the yard was designed to launch ships sideways into the narrow launching basin. Side-launching is not a great problem for small ships, but the C-3 type ships built by WPS were reaching the limit for this method. Risks to consider were: 1) healing over and foundering 2) damage to keel and rudder assembly 3) grounding or injury on the opposite side of the channel and 4) uncoordinated launching with bow or stem end leading. All these factors were considered and all forty-eight launchings went off with precision

USS Hamlin *ready for launching on March 5, 1942. Note foreward support for flight deck. Name was changed to HMS* Stalker *when she was lend-leased to Great Britain. Western Pipe & Steel photo No. 345, author's collection.*

Launching of the Sea Sparrow *on August 10, 1944. Smoke from the stack was to simulate boiler operation. Machinery was not yet operative. Western Pipe & Steel, author's collection.*

Ropes held the ship in place until launching when they were cut simultaneously with a guillotine apparatus. WPS News, author's collection.

and without any serious damage. The only damage found was the occasional bending of some bilge keels at the junction of the ship's bottom and side. In World War I, when shipbuilding and side-launchings were done at this same site, there were instances of damage to bilge keels.

The work of preparing a hull for launching was under the direction of the shipwright foreman, Jack Wilson. It was he who set the large wooden blocks on top of which the keel and ships bottom was placed. As the hull was completed and launching near, the hull was raised on a launching cradle. At this point the hull was held in place by several heavy ropes until the exact time of launching. As the ceremonies on the launching platform were completed and the sponsor was about to smash the champagne on the bow, the yard superintendent tipped his hat as a signal for launching to begin. A very important detail of the launching was the cutting of all the ropes at exactly the same instant so the ship would be free to slide sideways in its entire length. The cutting was done by a guillotine apparatus holding a twenty pound blade in a 350 pound chuck, permitting the blade to fall ten feet onto the rope to sever it. The eight guillotines were situated from bow to stern and electrically controlled under the direction of L.W. Delhi on the launching platform. The christening and the rope severance were thus controlled at the same point on the launching platform. In case of malfunction of the guillotine apparatus a man with an axe was stationed nearby to promptly cut the rope. But this was never necessary.

The launching was set at different times for each ship, and one launching took place at night. Launching was dependent on the tides and the amount of water in the launching basin to receive the ship. The WPS engineers noted early-on that the tide levels in the yard location were not the same as published for the San Francisco Bay in general. Tide rise and fall lagged significantly behind the San Francisco Bay figure. For this reason frequent checks of the tidal ebb and flow were made in the launching area so that the times for safest launching could be deduced. After careful engineering consideration, it was found best to have the water level one foot below the height of the sliding way. At this point the water level was 3.7 feet below the sliding way of ways 1 and 2 and 4.7 feet on ways 3 and 4. At this height the roll of the launched hull was minimized to about eight degrees and was relatively safe. The maximum roll noted was thirteen degrees on hull number 63.

At launching time the ship's stern was considerably heavier than the bow which could cause an uncoordinated launching with the stern

Left, W.H. Quarg, Asst. Regional Director of Finance, USMC, throws the switch that activates the guillotines. He is assisted by L.W. Delhi of WPS. WPS News, *author's collection.*

Clair Nooning, center, and her parents just before she launched the Sea Flyer. WPS News, *author's collection.*

entering the water first. To counter this problem, water was added in forward tanks or compartments to bring weight into balance.

In the tradition of shipbuilding each ship is christened by a lady, and this was always done at WPS. The sponsors of the forty-eight ships can be put into four classifications: 1) wives of famous people and politicians — Mrs. Leo Carrillo and Mrs. Earl Warren 2) workers at WPS — Mrs. Clarence Reiter and Clair Nooning and 3) friends or relatives of WPS executives — Miss Aileen Tallerday and Mrs T.S. Petersen and 4) wives of USMC officials Mrs Carl W. Flesher and Mrs. David Currier. See appendix three for a list of sponsors. The sponsor's husband was given the honor of throwing the electrical switch to activate the guillotines.

On several occasions the launching ceremony was used as a platform for war bond drives, speeches by war veterans or USMC officials and for presentation of awards. On at least two occasions the yard had war bond selling contests and the "queen" of the winning bond sales team had the privilege of christening a ship.

In the early part of the war a reception at the California Golf and Country Club followed the launching ceremony at the yard. The WPS and sponsor invited guests and presentations were made usually by Spike

Corbusier, publicist. The sponsor was given a gift such as a diamond or jewel encrusted watch or perhaps a silver tea set. The cost of these ceremonies of course was added to the ship's expenses and reimbursed by the USMC. In August 1943 an order came down from the USMC in Washington, D.C. that these frills were to be stopped.

Another launching tradition dictated that each member of the shipwright crew of a ship launching received a bright new silver dollar. This tradition was also followed at the WPS yard. In addition the husband of the sponsor for his part in pulling the switch for the launching guillotines, also was awarded a silver dollar since he was part of the launching crew.

As launching experience grew, more and more superstructure such as masts, deck houses and smoke stack were in place at launch time. On at least one occasion a smoke bomb was set off in the smoke stack so that it appeared the ship had steam up at launching.

Following launching into the narrow basin adjacent to the building ways, the hull was towed by tugs to the nearby outfitting docks. Some of the pilings for the outfitting docks are still present in 2000, sixty years after construction. Completion of the superstructure and machinery was

Just after the launching of the Steel Artisan *on September 17, 1941, the debris generated by the launching process is apparent. Western Pipe & Steel photo no. 279, author's collection.*

accomplished here. Due to overcrowding at the outfitting docks occasionally ships were towed to the San Francisco waterfront for outfitting there. The *Chandeleur* was outfitted at the repair dock run by Matson Navigation Company. The aircraft carrier HMS *Attacker* had the completion of her outfitting at a dock under the San Francisco Oakland Bay Bridge.

As each ship neared completion at the outfitting dock, a crew of divers came in to sweep her bottom This consisted of systematically inspecting the underwater hull and removing debris and wooden blocks from the building way which still clung to the ship's bottom. Otherwise such material could cause considerable damage to the propeller if it broke loose afterward.

At this point the ship left the SSF yard and was put into dry-dock for one or two days. Since WPS did not operate its own dry-dock, one was hired in the nearby Bay Area. Hunters Point Naval Shipyard, Bethlehem San Francisco Yard and Moore Drydock Company served this purpose at various times. For work in dry-dock the labor and staff of the dry-dock company were used under WPS supervision. The hull bottom was scraped and painted and careful inspection made for cracks in plates or welding

Above, the Sea Needle *and three other ships at the outfitting docks in May 1943. Right, the* Sea Thrush *begins the outfitting process in March 1945 and was completed that July. Western Pipe & Steel, above photo No. 644, both from author's collection.*

seams. In these inspections no damage was ever found to the keel or rudder resulting from the side-launching. The only damage recalled was some bending upward of a ten-inch angle iron situated at the junction of the ship's bottom and side. Repair of this damage was a minor task.

From dry-dock each ship was taken to a WPS leased dock on the San Francisco waterfront for completion. During wartime, WPS had ships at Piers 27, 88 and 90. However, Pier 27 was in continuous operation for WPS and was the site of the majority of WPS final outfitting. Some ships arrived in a near complete state and required only minor work plus final painting and clean-up prior to trials. Other ships, mainly the naval auxiliaries, i.e. seaplane tenders, destroyer tenders and escort aircraft carriers had considerable specialized equipment to install and spent more months in outfitting. In the case of the seaplane tender USS *Chandeleur*, which had one of the most protracted outfittings, a WPS crew of workers went over to Alameda Naval Air Station to complete work as the ship was being loaded with bombs, torpedoes and aviation fuel prior to her departure for the Pacific war zone.

With the completion of construction, each ship underwent a trial run. Each ship had a crew consisting of Captain Moore and licensed seamen.

U.S. Navy ships were taken on trial runs using the assigned navy crew. Henry Schoepp was hired as the chief engineer for WPS ships during their trial runs. Prior to the Pearl Harbor attack, trial cruises were made outside the Golden Gate in the Farallon Island area. After the war began, trials were done within the protection of San Francisco Bay antisubmarine nets as Japanese submarines were known to be in the area. All equipment was tested and checked off on a work list. Lists were make of any malfunctions or items not yet completed. Emphasis during the trial cruise was on the propulsion machinery, steering and communication equipment.

A fine description of a trial run was printed in the WPS News of April 1, 1945 when a WPS News reporter took the trial cruise of SS *Sea Blenny*. There the vessel was subjected to a three-hour endurance run at designed full speed to measure engine performance and fuel consumption. A further one-hour run followed, at the maximum possible ship's horsepower to test the ship's potential in case of emergency. Steering was tested at full speed. An emergency full astern stop was accomplished. This was one of the more dramatic tests as there was an abrupt change from full speed ahead to full speed astern. There was a half-hour test of running the ship stern first and steering astern at sixty rpms. Another test was to go full speed ahead from full speed astern.

At the close of the trial run a meeting was held in the ward room with representatives of WPS, the crew, and the operating company attending. Reliance was placed on one person from the operating department of the ship for operating performance and the chief engineer for machinery performance. At the close of this meeting any deficiencies were listed and a projection was made for the date of delivery of the ship to the operating company. Usual delivery was one or two days later. As the ship came back to Pier 27 a broom was usually seen flying at the masthead indicating a clean sweep of the trials. Very few of the WPS ships had to undergo a second trial run.

After completion the new ship was taken over by the new crew at the WPS dock and sailed into service. Occasionally ships came back for correction of defects or addition of equipment. The ships remained the property of the USMC and were operated by a shipping company which supplied the crew. Routing of the ship was under the control of the War Shipping Administration of the USMC. Naval ships were operated and crewed by the U.S. or British Navies.

CHAPTER 9
SHIPS FOR THE MILITARY

Escort Aircraft Carriers: Early in World War II, before Pearl Harbor, the German submarines and Focke-Wulf Condor aircraft were ravaging allied shipping in the Atlantic Ocean. Antisubmarine air patrols were based in the British Isles, Newfoundland, Canada and other British bases. But there remained a large gap in the central Atlantic where aircraft could not patrol and protect convoys. For this reason immediate aircraft interception of the long ranging Focke-Wulf Condor was needed. The answer seemed to be in aircraft operating from the convoy. Winston Churchill and Franklin Roosevelt took a personal interest in this problem and Roosevelt pressed Admiral Harold Stark, his chief of naval operations, for a solution involving conversion of merchant ships into auxiliary aircraft carriers. In March 1941 the first ship, the C-3 cargo-passenger ship *Mormacmail*, was acquired and converted in a mere three months. A similar prototype ship HMS *Archer* was converted for the British government at the same time. Based on the satisfactory performance of these ships, with some modification the C-3 type USMC hulls were deemed satisfactory for the proposed conversion. The Navy and USMC looked to the yards with as yet uncompleted C-3s for conversion to naval auxiliaries. WPS was one of the few yards constructing C-3s at the time. In December 1941 the Navy signed contracts with several firms to convert twenty-four C-3 hulls to aircraft carriers. WPS was among those chosen and assigned to convert four hulls to this purpose. The first of these four ships was the SS *Steel Artisan*, WPS hull number 62, then nearing completion at the outfitting docks of the SSF yard. It was necessary to clear away all deck houses and other superstructure down to the main deck. While this was being done the shops were manufacturing the sponsons and deck supports to be placed across the ship to support the flight deck. With the many complex engineering and manufacturing problems involved, it took nine months work to complete the ship and deliver her to her British crew as HMS *Attacker*.

HMS Attacker, *lower right, at San Francisco's Embarcadero, was a WPS C-3 converted from cargo ship to escort carrier. The Fairey Swordfish, upper left, were British planes of the type carried aboard such carriers. FAA Museum, author's collection.*

The next three aircraft carriers, HMS *Stalker*, HMS *Fencer* and HMS *Striker*, were converted while still on the ways before launching. Some of these ships even had a few of the deck supports in place at the time of launching. These ships also took nine to ten months to complete after launching with delivery taking place December 1942, February 1943 and April 1943. Carl W. Flesher, Pacific Coast Director of Construction USMC, in a speech at the WPS yard on Maritime Day May 23, 1943 took note of some of the intricacies of the aircraft carrier construction. The complexity of the conversion required 36,000 blueprints and 80,000 purchase orders. The installation of catapults, large aircraft elevators, arresting gear, radar, guns, fire-fighting equipment and special sprinkling systems each required special design and installation. Crew accommodations were increased to 850 men. A special island superstructure was created and placed on the starboard side of the flight deck to contain the navigation bridge and captain's sea quarters. Special ready and briefing rooms were provided for flight crews. Offices for officers of the flight squadron as well as the ship's crew were provided. Special shops for engine and aircraft repair and maintenance were incorporated and a special magazine area provided for stowage of bombs, torpedoes, depth

Here can be seen the vastness of an escort carrier's hanger deck. Western Pipe & Steel, author's collection.

charges, and fuel tanks for high octane aviation fuel. The feeding and medical and dental requirements for such a crew were also included. In all, each carrier required 2,800,000 man-hours or the equivalent of the time taken to build three C-3 cargo ships.

After placing the transverse deck supports a steel deck of one-fourth inch plates was installed. On top of the steel deck were bolted planks of selected Oregon fir for the planes to land upon. The use of a wood landing surface was standard in the U.S. Navy whereas in the British Royal Navy a steel deck was customary. In the case of these carriers the American type of deck was used in spite of their intended use by Great Britain under the Lend-Lease Agreement. The caulking of the deck timbers became a special problem to the yard. WPS had no expert caulkers and had to locate men with this dying skill. Eventually they were found among former shipyard workers who had worked on wooden ships. Many of these men were old and could work only half days, but were recruited for this special task and to train others in this craft.

The aircraft carriers were moved to Pier 90 in San Francisco for installation of many specialty items supplied by the Navy and installed by WPS men under Navy supervision. Examples of such equipment were the guns, radar and communication equipment. The catapult to be used for launching aircraft was tested by a special trolley weighing as much as a plane. A cable was attached to the trolley so that after launching into the Bay over the bow, it could be reeled back onto the ship.

During the later phases of outfitting the British crew arrived by train from the East Coast. They were housed on the ship and many are the tales of parties they had aboard. One consistent observation among workers was of the extreme youth and inexperience of most of these crews. The captain, especially of HMS *Attacker*, was experienced and knew what he wanted. Shortly after arrival he let it be known that he wanted the shower removed from the captain's cabin and replaced by a bathtub. This the outfitting dock supervisor said he could not do; however, the captain went immediately to the U.S. Navy and soon a work order came through to put in the tub.

Two trial runs were made using the British crews. The British crews seemed to like San Francisco and found more and more "defects" which had the effect of delaying their departure for the East Coast and England. One of the last of the trivial defects was the ridging along the welded plate seams of the floor of the hanger deck. After receiving this complaint

The young age of the British crew and the complexity of an escort carrier are apparent in this stern view of the HMS Attacker *at San Francisco. Orrie Damewood, author's collection.*

chippers with their noisy pneumatic chipping apparatus were put to work during the graveyard shift. Of course, this resulted in no sleep for the British crew and the chipping did not have to continue long.

After acceptance, the aircraft carriers each set out via the Panama Canal for the East Coast. HMS *Attacker* carried several Swordfish aircraft for anti-submarine patrol. These planes had flown overland to Alameda Naval Air Station from Halifax, Nova Scotia. No airplanes or air crews were on the other ships as they sailed. A rumor spread around the WPS yard that one of the carriers was sunk by a submarine and was lost with all hands before reaching England. This a WPS worker told me over forty years later, and he still believed it at that time. This story is a testimony to strict wartime news security, but in this case the story was not true, and it must have had some demoralizing effect on the men working so hard. The record of these four ships great contribution to the war will be covered later, but their operating areas included the Atlantic, Arctic, Indian and Pacific Oceans and the Mediterranean Sea.

Troop Transports: Prominent among the many uses the adaptable C-3 hulls could be put to was the troop transport. With their steam turbine-driven speed of eighteen knots they had better protection for the troops against submarine and air attack than the slower reciprocating-engined Liberty ships. The C-3's also had the substantial capacity needed for the troops and their field equipment.

Twenty-three of the WPS ships were completed as attack troop transports — fourteen for the Navy and nine for the Army. In addition five more WPS ships were converted to troop transports elsewhere after they had completed at least one wartime cruise as a cargo carrier. These conversions were carried out by the Moore Drydock Company in Oakland and the Marine Repair Shop operated by the San Francisco Port of Embarkation. Some ships went to the East Coast for conversion to APAs (attack transports). The troop capacity ranged from 1906 to 2838 with most carrying about 2100 men. In addition to the troops, the ships had about 250,000 cubic feet of cargo space for equipment.

The hull and contour of the superstructure was similar to the cargo ship version. The main difference in external appearance was the presence of landing craft in special davits on deck. Two holds amidships were converted for troop use. Three deck levels were created in these holds. On each level bunks tiered five high were constructed in rows. Each

Typical of the C-3s converted to troopships was the Sea Sturgeon, *here shown after months of hard use. National Archives.*

passageway had bunks on either side. Bunks were thirty inches wide. A galley and mess hall separate from the ship's crew was provided for the troops. The troops used their own mess gear and had no assigned dining room. A sick bay and dental clinic were provided for the troops. The troop commander had a private cabin and office with a loudspeaker communication to all troop areas from either location. The communications system in these ships was quite complex. The ship's captain had direct communication with all compartments in case of emergency. He had direct separate contact with the ship's operating departments. To entertain troops on their long crowded voyage, there were facilities to broadcast music and other entertainment to troop compartments.

Since the troopships carried less weight, concrete was placed in the bottom of the holds as ballast to improve the ship's comfort, safety and performance.

Some troopships were operated by the Navy, others by the Army Quartermaster Corps and still others by the War Shipping Administration.

CHAPTER 10
SHIP REPAIR WORK

The last USMC contract for construction of C-3 type ships was awarded in the early war years. During the succeeding three years most of the ships were finished and the last six were on the ways and fitting out. The Normandy Invasion in Europe was a proven success and the Allied Armies were overrunning Europe. New ship construction was winding down, and there was little prospect for further contracts. Rumors of the yard closing were rampant and some employees of WPS began looking for work elsewhere.

In October 1944 came word that the Navy was working out a contract with WPS for ship repair work. Of course, WPS without a dry-dock, could not undertake some kinds of repair, but they had always shown ability in adapting the C-3 type ships for many uses and their skills were respected. As the war in the Pacific escalated, many ships sustained damage from gunfire and Kamikaze attacks, particularly in the Okinawa invasion. The redeployment of all forces to the Pacific region, with the ultimate plan of landings and battle on the Japanese home islands, would greatly increase the number of ships in the Pacific. In January 1945 WPS received contracts from the Navy and War Shipping Administration for ship repair work. A similar contract with the Army was expected shortly thereafter. The repair work was to be done mainly at Piers 27 and 88 and elsewhere along the San Francisco waterfront. Some of the repair work would be done at the SSF yard, but modifications to the yard were needed.

The *WPS News* of March 1, 1945 gave projected manpower needs for WPS repair work as determined by the Navy's Bureau of Ships. Beginning with 1500 workers in March 1945, it was expected that 3000 would be needed by June 1945. The first repair job started on January 8, 1945 at Pier 27. By March 1945 there were three WPS contracted ships tied up at Pier 27 for repairs. The volume of repair work fluctuated, but again in the April 15, 1945 *WPS News,* activity at the docks had increased

Before, above, and after repairs to the Liberty ship Philip C. Shera. WPS News, *author's collection.*

with three major repair jobs in hand. One of the largest was on the liberty ship *Phillip C. Shera* which collided with a tanker off the Golden Gate. She had a fifty-foot-long gash in the port bow extending from the second deck to the main deck. Repair was accomplished at Pier 88 in one month. The chief steward later reported that the repaired galley was better than ever.

In May 1945 there were fourteen ships at Pier 27 awaiting repair. By mid-May *WPS News* reported forty-two ship repair jobs completed. Seven hundred and fifty men were working on repairs on the San Francisco docks. Also in May the first two ships came into the SSF yard for repair in the launching basin, the SS *Jane Adams* and USS *Rutilicus*.

In July 1945, as reported by the *WPS News*, two new jobs came in. One was the *LST 599* which was hit by a Kamikaze that crashed through the main deck and ended up in the tank deck. The LST captain obtained the pilot's family belt as a war souvenir. The other job at the yard was USS *Callaway*. The *Callaway* started life as WPS hull number 92, having been built by WPS in SSF. After eighteen months of continuous service at sea in the Pacific area she returned for refit and modification.

In the summer of 1945 due to the increasing activity at Pier 27 the personnel and engineering offices were remodeled and enlarged.

Upon completion of shipbuilding contracts in 1945, the facilities were used for ship repair. The SS Sea Cardinal *in foreground was built by Western Pipe & Steel. Orrie Damewood photo, author's collection.*

The August 1, 1945 issue of *WPS News* noted that 118 ships had been repaired in San Francisco and SSF. Subsequently a published breakdown of repair work indicated work on 83 ships at Pier 27 and ten at the main yard. Presumably the other twenty-one jobs were at other piers such as Pier 88. Pier 39 on the San Francisco docks was also used by WPS for ship repair.

The time for repair work averaged ten days per ship so that, on average, one job was completed every two and one-half days.

Then came the unexpectedly early surrender of Japan following the atomic bomb explosions. There were still ships requiring both repair of battle damage and routine overhaul, but by September the pace was slackening. Many employees left to find peacetime work elsewhere. A complete list of the ships repaired by WPS has not been found by the author, but the following list shows some of the repair jobs and the location of the job:

SS *Phillip C. Shera*, liberty ship	Pier 88
SS *Jane Adams*	SSF
USS *Rutilicus*, cargo ship	SSF

LST 599, tank landing ship	SSF
USS *Callaway*, troop transport	SSF
USS *Barrow*, troop transport	Pier 27
SS *Sea Cardinal*, cargo ship	SSF
USS *Berrien*, troop transport	?
USS *Admiral H.T. Mayo*, troop transport	Pier 27
USA *Comfort*, hospital ship	SF

The end of the World War II shipbuilding was at hand. There were far more ships than were needed for the peacetime economy. It was easy to predict difficult times for the United States shipbuilding industry!

Chapter 11
Post World War II

With World War II under control and winding down by 1944, new contracts from the USMC for merchant ships were scarce. Greater emphasis was placed on special needs such as landing craft and destroyer escorts. With the termination of the war in 1945 unbuilt ships were cancelled.

As we have seen, many shipyards such as WPS turned for a time to ship repair. Late in the war as a result of Kamikaze attacks and close inshore landing operations there had been increased need for more ship repair facilities; however, even this work ceased with the war's end and the yards with facilities better suited to ship repair work dominated this field.

In 1943 deterioration caused by rotting and infestation of timbers and pilings was noticed in the facilities at the California Shipbuilding Corporation. Based on this the USMC requested that other shipyards check their facilities. In September 1944 a survey was taken at WPS which showed considerable damage from rot and marine borers. The damage was more advanced in ways 1 and 2 which were constructed at an earlier date in 1939; ways 3 and 4 and the outfitting docks were sound. Extensive repairs to ways 1 and 2 were recommended to make the structures safe for major load-bearing. By late 1944, however, the contracted C-3 construction was coming to an end and there was no immediate prospect for a new contract. The last ship built on way number 1 was launched in March 1945 and the last ship on way number 2 was launched in May 1945. This ended WPS ship construction. Repairs to the ways were not undertaken. The deterioration did mean that should future contracts be offered, a major repair of ways 1 and 2 would be needed.

Shortly after the end of World War II, WPS was sold to the Consolidated Steel Corporation and the new entity of Consolidated Western Steel Corporation was created.

In September 1945 a survey of the SSF plant was made by the Consolidated Steel Corporation which had purchased WPS. This survey showed a total of 1,117,383 square feet of building space. Of this 814,158 square feet were owned by Consolidated Steel Corporation and 303,225 square feet were owned by USMC. These figures included shops, warehouses, administration buildings, cafeteria and hospital.

In May 1948 Consolidated Western Steel Corporation completed a questionnaire on the status of their shipbuilding facilities at the SSF plant. The four building ways were in poor condition. Underground piping and outside equipment showed increasing deterioration. However, trackage, shops and tools remained in good condition.

In 1949 the War Munitions Board in Washington, D.C. began provisional plans in case a national emergency developed. This Board inquired as to the suitability of the SSF shipyard for new construction or repair work. Correspondence on this matter was carried out through the headquarters of United States Steel Corporation in Pittsburgh and headquarters of Consolidated Western Steel Corporation, a subsidiary, in Los Angeles. It was felt that the ship repair capabilities were quite limited at SSF because the plant lacked a dry-dock and had limited machine shop facilities capable of major machinery repairs. It was thus concluded that ship repair work was not viable.

Attention then was turned to the possibility of ship conversion and new construction. For these types of work the SSF yard was more suited; however, the rundown state of the building ways and the silted up channel were a concern. An extensive revision, including filling in the launching channel and rebuilding the ship-ways with four end-launching ways and dredging the channel into the yard and turning basin to an eighteen-foot depth, would be needed for the yard to become useful in conversion and new construction.

In 1949, the Navy Bureau of Ships tentatively allocated construction of sixteen attack cargo ships (AKA) to the SSF yard to be completed over a three-year period, providing a national emergency developed.

In 1950 the intensity of planning in Washington, D.C. by the Munitions Board was increased because of escalating conditions in the Far East, particularly the Korean situation. Estimates of $150,000 were obtained for dredging the channel and basin. In addition, $475,000 would be required to rehabilitate and equip shops, docks and yard facilities for ship repair work. This figure did not include rebuilding the ways.

Then, in September 1952 word was received from the U.S. Navy that the emergency plans for the SSF shipyard had been changed and the construction of attack cargo ships was no longer needed. In place of the AKAs the yard was assigned to build landing ships dock (LSD). The proposed schedule called for construction of twelve LSD's over a three-year period. At the time the prototype of these ships, LSD-28, was under construction by the Ingalls Shipbuilding Corporation of Pascagoula, Mississippi.

Considerable correspondence developed between Ingalls and Consolidated Western Steel over plans to build LSD's. Ingalls officials visited SSF on several occasions and looked favorably on including the SSF yard in the LSD building program. Again the question of side-launchings arose and the Ingalls people advised reconstruction of the SSF yard for end-launching. In February 1953 a survey of the SSF yard indicated it would cost about seven million dollars to rehabilitate and re-equip the yard providing side-launching was retained as in the present yard plan. A reconstruction of the yard for end-launching would cost considerably more. In fact, an additional estimate was obtained in April 1954 for a completely rebuilt yard with end-launching facilities. This estimate was $11,228,000. The great expense of rehabilitating the yard and converting to end-launching continued to hold the SSF yard back. Any plans for the rebuilding of the yard would depend on the occurrence of a national emergency. Yet the Korean war came and went without the need to develop the SSF yard.

The Ingalls Shipbuilding Corporation completed their initial contract for six LSDs, however Consolidated Western Steel Corporation received no contracts. By 1956 correspondence indicates that the tentative emergency planning for the SSF yard was no longer for LSD's but for a different type of ship. The new type of ship was not specified in the correspondence. This then was the end of consideration of building ships at the SSF plant.

The Consolidated Western Steel Corporation plant in SSF was taken over by the United States Steel Corporation's American Bridge Division in 1949. Only one significant maritime event occurred during the U.S. Steel tenure. In the early 1970s Howard Hughes' Summa Corporation began construction of the *Glomar Explorer* and the large submersible barge *HMB-1* in a top-secret salvage attempt to recover a Soviet nuclear submarine which had sunk in the mid-Pacific. The barge was equipped with a grappling mechanism made of a special extra-strong steel alloy. *HMB-1* came from her home base in Redwood City, California to the U.S. Steel plant in SSF. According to employees at the time it was necessary to dredge the channel to the plant and to remove a corner of the shop building to accomplish the work. The entire submarine salvage operation, the Jennifer project, was a highly classified secret, and even today, twenty-five years later, details are difficult to obtain.

CHAPTER 12
CONCLUSION

In closing, it is interesting to take note of the final disposition of this shipyard and its ships. None of the ships remain afloat. All have been scrapped except the six which, as casualties of war, lie at the ocean's bottom. One of these was sunk in the service of Italy, the victim of a British surface force from Malta.

The shipyard itself has been out of service since 1946. The land now is being redeveloped into commercial property. There will never be another shipyard at this South San Francisco location.

PART II
THE SHIPS

Isanti

WPS Hull No.1.
Tonnage:

Gross	5713
Net	4129
Deadweight	5726

Length: 410.5 feet
Beam: 54 feet
Freeboard: 5 feet 9 1/2 inches
Draft: 24 feet 2 inches
Engine:

General Electric steam turbine
2500 SHP
Oil fuel

Speed: 11 knots
Two decks
Crew: 39
Date launched: June 2, 1918.

Departed San Francisco October 6, 1918 for San Pedro and New York. She made numerous trips between East and Gulf Coast U.S. ports and South American ports to ports in Germany, Belgium, France, and England. Last trip completed on arrival in Philadelphia October 22, 1921. During this last trip she put into Halifax, Nova Scotia and St. Johns, Newfoundland for repairs. On October 24, 1921 *Isanti* was transferred to the reserve fleet and tied up at Hog Island in Group 10B. She was owned by the United States Shipping Board during her entire operating history.

Disposal: On June 27, 1930 the *Isanti* was sold to the Union Shipbuilding Company of Fairfield Maryland for scrapping.

Patriotism is apparant on the beam being lifted aboard the Isanti *before launching. Modern safety rules would never allow a worker to ride the beam as he is in this photo. Western Pipe & Steel photo no. 144, author's collection.*

Nantahala

WPS Hull No.2
Tonnage:

Gross	5714
Net	4045
Deadweight	8726

Length: 410.5 feet
Beam: 54 feet;
Freeboard: 5 feet 9 1/2 inches
Draft: 24 feet 2 inches;
Engine:

General Electric steam turbine
2500 SHP
Oil fuel

Two decks
Crew ?
Date Delivered: October 31, 1918.

Departed San Francisco December 5, 1918 for New York via Panama Canal arriving December 31, 1918. Went into service between ports on East Coast of USA and South America to European ports in Italy, Spain, England and Holland. Put in for repairs: New York, April 1919; New York, November 1919; New York, July 1920; New York, May 1921. Held survey August 26, 1921. Moved to Staten Island Reserve Fleet off Totenville, New York in Section 2 September 2, 1921. *Nantahala* was owned by the United States Shipping Board during her entire operating history.

Disposal: On October 30, 1929 *Nantahala* was delivered at New York to the Union Shipbuilding Co. of Fairfield, Maryland for scrapping.

Moments after launching the Nantahala's wave can be seen cresting across the basin, right. Western Pipe & Steel photo no. 88, author's collection.

West Avenal

WPS Hull No.3.
Tonnage:
    Gross   5692
    Net   4020
    Deadweight   8735
Length: 410.5 feet
Beam: 54 feet
Freeboard: 5 feet 9 1/2 inches
Draft: 24 feet 2 inches
Engine:
    General Electric steam turbine
    2500 SHP
    Oil fuel
Two decks
Crew 46.
Date Delivered: February 1, 1919.

Departed San Francisco with cargo for Food Administration February 15, 1919 arriving Norfolk April 4. Departed Norfolk for Leith, Scotland on April 9, 1919. Had to return because of engine breakdown. Remained in service between East Coast and South American ports and European ports in Spain, France, and England. In January 1920 she was beached on Staten Island because she was threatening to sink. Repaired from January 15 to February 27, 1920. On August 19 *West Avenal* received minor damage in a collision at New York. Underwent repairs in New York January 6 to 25, 1921 following which she was tied up in the Staten Island Sound Reserve Fleet 3 Section 1. *West Avenal* remained under the ownership of the United States Shipping Board for her entire life.

Disposal: On August 29, 1929 at noon *West Avenal* was delivered to the Union Shipbuilding Co. of Fairfield, Maryland in New York for scrapping.

Oskaloosa

WPS Hull No.4.
Tonnage:

Gross	5623
Net	4189
Deadweight	8750

Length: 410.5 feet
Beam: 54 feet
Freeboard: 5 feet 9 1/2 inches
Draft: 24 feet 2 inches
Engine:
General Electric steam turbine
2560 SHP
Oil fuel
Two decks
Crew 44
Date Delivered: December 20, 1918.

Departed San Francisco January 11, 1919 after loading at Port Costa, California. Arrived New York via Panama Canal. Went in service between East and Gulf coasts of U.S.A. and European ports in England, France, and Belgium. Damaged in collision with SS *Winneconne* in April 1921. Required repeated repairs in: New Orleans, March 1920; Norfolk, September 1920; New Orleans, January 1921; Jacksonville, July 1921. On July 4, 1921 she put out a distress call and had to be towed to the Azores with disabled engines. In September she again had engine trouble requiring a tow by SS *Monroe* into St. Johns, Newfoundland. Then towed from Newfoundland to Boston. Subsequently was towed to Norfolk by tug *Butterfield.* Put into reserve fleet at Camp Eustis, Hampton Roads on July 14, 1922. *Oskaloosa* remained under ownership of United States Shipping Board for her entire life.

Disposal: *Oskaloosa* was delivered to the Union Shipbuilding Co. of Fairfield, Maryland at Norfolk, Virginia for scrapping on July 19, 1929.

Stern view of the Oskalousa *while still under construction. Western Pipe & Steel photo no. 63, author's collection.*

West Vaca

WPS Hull No.5.
Tonnage:

Gross	5548
Net	4151
Deadweight:	8800

Length: 410.5 feet
Beam: 54 feet
Freeboard: 5 feet 9 1/2 inches
Draft: 24 feet 2 inches
Engine:

General Electric steam turbine
2500 SHP
Oil fuel

Two decks
Crew 42.
Date Delivered: March 27, 1919.

Departed San Francisco April 4, 1919 for Yokohama, Shanghai, Manila and Hong Kong, returning to Seattle July 23. Departed again for Manila, Hong Kong and Shanghai returning to San Francisco October 30, 1919. Departed San Francisco December 26 for New York and Philadelphia. On the return trip starting March 13, 1920 she charted a course through the Panama Canal to Honolulu, Yokohama, Kobe, Miiki, Manila Singapore, Batavia, Suez, Breast and London. On route from London to Boston she collided with schooner *John Gibson* off Tilbury on November 10, 1920. *West Vaca* arrived New York December 30, 1920 and was tied up in the Staten Island Reserve Fleet 1 Section 3. *West Vaca* remained under the ownership of the United States Shipping Board for her entire life.

Disposal: *West Vaca* was delivered to the Union Shipbuilding Co. of Fairfield, Maryland at New York at 10:20 A.M. on August 23, 1929 for scrapping.

West Ashawa

WPS Hull No.6.
Tonnage:

Gross	5609
Net	4095
Deadweight	8750

Length: 410.5 feet;
Beam: 54 feet
Freeboard: 5 feet 9 1/2 inches
Draft: 24 feet 2 inches
Engine:

General Electric steam turbine
2800 SHP
Oil fuel

Two decks
Crew 49.
Date Delivered: May 21, 1919.

Departed San Francisco for Portland May 22, 1919 and thence to New York arriving July 4, 1919. In service between East Coast and Gulf ports of U.S.A. and European ports in England, Holland, Spain and France. Put into Hampton Roads for repair February 1921. Shifted to reserve fleet at Camp Eustis March 29, 1921. Out for repairs July to September 1921 and then back to Unit 25 at Camp Eustis-Hampton Roads Reserve Fleet. *West Ashawa* remained under ownership of United States Shipping Board during her entire life.

Disposal: *West Ashawa* was delivered at Norfolk to the Union Shipbuilding Co. of Fairfield, Maryland for scrapping.

The wake from the stern of the tug on the West Ashawa*'s starboard bow would indicate she is pushing, probably maneuvering the ship toward a berth. Photograph Courtesy Peabody Essex Museum, neg. V-1321.*

West Alcoz

WPS Hull No.7.
Tonnage:

Gross	5606
Net	4230
deadweight	8800

Length: 410.1 feet
Beam: 54 feet
Freeboard: 5 feet 9 1/2 inches
Draft: 24 feet 1 1/2 inches
Engine:

General Electric steam turbine
2500 SHP
Oil fuel

Two Decks
Crew 40.
Date Delivered: June 24, 1919.

Departed San Francisco for New York via Honolulu on June 26, 1919. Made four round trips between East Coast and Gulf ports to European ports in England, Belgium and France between September 1919 and January 1921. On last trip *West Alcoz* put into New London for fuel and repairs in January 1921. Shifted to Reserve Fleet at Camp Eustis-Hampton Roads Unit 2 on August 18, 1921.

West Alcoz remained under ownership of United States Shipping Board during her entire life.

Disposal: *West Alcoz* was delivered to purchaser Union Shipbuilding Co. of Fairfield, Maryland at Norfolk at 11:00 AM on July 9, 1929 for scrapping.

West Aleta

WPS Hull No.8.
Tonnage:
    Gross   5605
    Net   4337
    Deadweight   8860
Length: 410.5 feet
Beam: 54 feet
Freeboard: 5 feet 9 1/2 inches
Draft: 24 feet 1 1/2 inches
Engine:
    General Electric steam turbine
    2500 SHP
    Oil fuel
Two decks
Crew 42.
date Delivered: August 1919

Departed San Francisco for Hull, England on August 10, 1919 with cargo of grain arriving September 14. She returned to San Francisco via Norfolk, Bremerton, and Seattle and was drydocked in Seattle. In San Francisco she had a cracked turbine repaired. On January 6, 1920 she commenced a voyage to Hamburg arriving Cristobal on January 21. On February 13 she was reported stranded in breakers seven miles west north west of Brandaris light on Terschelling Island. *West Aleta* subsequently broke up and on June 10, 1920 was reported a total loss.

The launching of the West Aleta *in 1919. Note the observers just beyond the bow leaning with the ship as she slides in the water. San Francisco National Historical Park, P83-1042.1x.*

West Cactus

WPS Hull No.9.
Tonnage:

Gross	5643
Net	3523
Deadweight	8800

Length: 410.5 feet
Beam: 54 feet
Freeboard: 5 feet 9 1/2 inches
Draft: 24 feet 2 inches
Engine:

Joshua Hendy Iron Works 3-cylinder
   triple expansion
2800 IHP
Oil fuel

Two decks
Crew 44.
Date: Delivered: August 1919.

Departed San Francisco September 9, 1919 for Yokohama via Honolulu for engine repair. *West Cactus* made four round trips between San Francisco and Asiatic ports of Yokohama, Shanghai, Manila, Kobe, Singapore, Samaraz, Tjilatjap, Saigon and Hong Kong between September 1919 and February 1921. She was laid up in February 1921 until April 1923. There were two trips to the Far East between April and December 1923. Then followed two round trips to the South American ports of Montevideo, Santos, San Juan, Cartagena and Bahia between March 1924 and February 1925. In September 1925 there was another Far East trip followed by two trips to South America returning to San Francisco in late July 1926. During this period *West Cactus* remained the property of the United States Shipping Board.

On September 16, 1926 *West Cactus* was sold to the Pacific Argentine Brazil Line Inc. at San Francisco. She remained in this service to South America with San Francisco as home port until 1942. In 1942 ownership was passed to Pope & Talbot Inc. of San Francisco with San Francisco remaining the home port. In 1945 ownership changed to the War Shipping Administration, Washington D.C.

Disposal: *West Cactus* was out of service between 1945 and 1947 at which time she was sold to the American Iron & Metal Co. of Emeryville, California for scrapping.

The West Cactus *with a deck load of lumber. Photograph Courtesy Peabody Essex Museum, neg. H-2101.*

West Caddoa

WPS Hull No.10.
Tonnage:

Gross	5641
Net	3522
Deadweight	8800

Length: 410.5 feet
Beam: 54 feet
Freeboard: 5 feet 9 1/2 inches
Draft: 24 feet 2 inches
Engine:

Joshua Hendy Iron Works 3-cylinder
triple expansion
2800 IHP
Oil fuel

Two decks
Crew 45.
Date Delivered: September 10, 1919.

Departed San Francisco for Yokohama on her maiden voyage arriving December 6. In 1919 and 1920 there were four round trips to the Orient followed by a fifth where the return was via the Mediterranean Sea and East Coast of U.S.A. From 1922 to 1930 she was in service between the U.S.A. Gulf Coast and England with occasional stops at Continental ports.

There is no record of sailing activity or reserve status from 1930 to 1940, but *West Caddoa* did remain under ownership of the United States Shipping Board and its sucessor the United States Maritime Commission until 1940. In 1940 ownership was transferred to Ministry of War Transport of the British Government. She was renamed *Empire Guillemot*. On October 24, 1941 she was torpedoed and sunk by aircraft off Galeta Island in the Mediterranean Sea.

The West Caddoa *with tug alongside, probably entering port. Photograph Courtesy Peabody Essex Museum, neg. V-1460.*

West Kader

WPS Hull No.11.
Tonnage:
Gross	5570
Net	3468
Deadweight	8800

Length: 410.5 feet
Beam: 54 feet
Freeboard: 5 feet 9 1/2 inches
Draft: 24 feet 1 1/2 inches
Engine:
   Joshua Hendy Iron Works 3-cylinder
     triple expansion
   2800 IHP
   Oil fuel
Two decks
Crew 44
Date Launched: July 2, 1919.
Date Delivered: December 31, 1919.

Departed San Francisco on maiden voyage January 8, 1920 for the Far East via Portland, Oregon. On return she made a trip to Havana and returned, and then in August 1920 made a trip to Cork, Ireland. *West Kader* then returned to Portland via Norfolk and Honolulu. Beginning in April 1920 until April 1928 she remained in service between Portland, Oregon and Far East ports in China, Japan, Russia, Philippines, and Hong Kong.

At 5:00 P.M. on June 4, 1928 *West Kader* was delivered by the United States Shipping Board to the purchaser States Steamship Company at Portland, Oregon. The ship was renamed *New York* and her homeport remained Portland. In 1937 ownership was transferred to the Everett Steamship Company of Mobile, Alabama with the new home port of Mobile, Alabama. The ship was renamed *Pan Kraft*. In 1939 Pan Atlantic Steamship Corporation was listed as owner and homeport remained Mobile. Waterman Steamship Company became managers for *Pan Kraft*. In 1939 her home port became Wilmington, Delaware.

In June 1942 *Pan Kraft* started on a voyage from New York to Archangel, U.S.S.R. via Iceland with a load of crated aircraft and a deck load of planes. She became part of Convoy PQ-17 which subsequently was dispersed when threatened by the possible presence of the German battleship *Tirpitz*. *Pan Kraft* was attacked and bombed by Junkers 88 bombers in the Barents Sea on July 5, 1942. Bomb explosions or near misses resulted in ruptured oil and steam lines and the ship was abandoned. The convoy escort HMS *Lotus* then tried to sink *Pan Kraft* by shelling with deck guns, but without success. Sinking finally occurred at 6 A.M. on July 7, 1942 after an explosion.

The West Kader *with a full load of lumber at Portland, Oregon. Oregon Historical Society, neg. Gi10756.*

West Cadron

WPS Hull No.12.
Tonnage:

Gross	5724
Net	3564
Deadweight	8660

Length: 410.5 feet
Beam: 54 feet
Freeboard: 5 feet 9 1/2 inches
Draft: 24 feet 2 inches
Engine:

Joshua Hendy Iron Works 3-cylinder
    triple expansion
2800 IHP
Oil fuel

Two decks
Crew 44.
Date Delivered: March 3, 1920.

Departed San Francisco on maiden voyage to the Orient on March 12, 1920, visiting Kobe, Shanghai, Manila and Hong Kong before returning to San Francisco June 17th. She made three round trips from San Francisco to the Far East and then was laid up in reserve on January 21, 1921 at Southampton Bay. Transferred to Laidup Fleet Comm. February 16, 1923.

After reactivation the *West Cadron* went into service between the Pacific Northwest ports of Everett, Seattle, Tacoma and Portland and the Orient. She remained in this service from June 1923 until August 1928.

On August 11, 1928 the *West Cadron* was delivered by the United States Shipping Board to the purchaser States Steamship Company at Portland, Oregon. She was renamed SS *Iowa*. The *Iowa* continued in service of States Steamship Company for the rest of her life. She was lost with all thirty-four hands when she foundered on Peacock Spit near Cape Disappointment at the mouth of the Columbia River on January 12, 1936.

The wreck of the Iowa, *ex-*West Cadron, *as viewed from a rescue vessel. Columbia River Maritime Museum.*

West Cahokia

WPS Hull No.13.
Tonnage:

Gross	5645
Net	3500
Deadweight	8800

Length: 410.5 feet
Beam: 54 feet
Freeboard: 5 feet 9 1/2 inches
Draft: 24 feet 2 inches
Engine:

Joshua Hendy Iron Works 3-cylinder
   triple expansion
2800 IHP
Oil fue

Three decks
Crew 44.
Date Delivered: April 16, 1920.

Departed San Francisco April 18, 1920 for maiden voyage to Portland, Shanghai, Iloilo and Manila. On return passage she went to Philadelphia via Honolulu and Panama. The *West Cahokia* stayed in service between the East and West Coasts of U.S.A. and Europe. She was laid up at Camp Eustis, Hampton Roads September 7, 1921. Between 1922 and December 1925 *West Cahokia* was in service between San Francisco, Northwest Pacific ports and the Orient.

On March 26, 1926 the *West Cahokia* was sold by the United States Shipping Board to Ocean Transportation Company of San Francisco and was renamed *Pacific Fir*. In 1928 she was sold to Dimon Steamship Corporation and New York became her home port. In 1933 the ship was lent to Rear Admiral Richard E. Byrd and was renamed *Jacob Ruppert*. In 1936 ownership reverted to the U.S. Maritime Commission, Washington, D.C. with homeport of New York. Ownership remained with the USMC until 1941 when she was sold to the North Atlantic Transport Company of Panama her homeport was changed to Panama, R.P. Her name was changed to *Cocle*. On May 5, 1942 *Cocle* was torpedoed and sunk by a German submarine north of the Azores.

Launching of the West Cahokia *was an occasion for flags and pagentry. San Mateo County Historical Association.*

West Calera

WPS Hull No.14
Tonnage:

Gross	5644
Net	3524
Deadweight	8800

Length: 410.5 feet
Beam: 54 feet
Freeboard: 5 feet 9 1/2 inches
Draft: 24 feet 2 inches
Engine:

Joshua Hendy Iron Works 3-cylinder
   triple expansion
2800 IHP
Oil fuel

Two decks
Crew 45
Date Delivered: January 14, 1920.

Departed from San Francisco on maiden trip to San Pedro June 11, 1920. Departed San Francisco August 19, 1920 for the Orient returning via Calcutta, Columbo, Port Said, Bizerte, Marseilles, Bologne, New York, Baltimore and Panama. Departed again for the Orient via Los Angeles returning to San Francisco December 15, 1921. The *West Calera* was transferred to Southampton Bay January 13, 1922. She was transferred to Laid up Fleet Comm. March 18, 1923 and again transferred from Laid up Fleet Comm. to Operating Dept. later in 1923. On return to active service *West Calera* made two trips to the East Coast of South America and then went into service between West Coast U.S.A. and New Zealand, Australia and the Far East from November 1924 until June 1928.

On July 5, 1928 at 10:30 A.M. *West Calera* was sold and delivered to the purchaser Oceanic and Oriental Navigation Company Inc. of New York. She was then renamed *Golden Harvest.* Her port of registry was San Francisco. She remained in this service until 1937 when she was sold to American-Hawaiian Steamship Company of New York and renamed *Oklahoman.* Port of registry became New York. The *Oklahoman* stayed in service with American-Hawaiian for the rest of her life.

On July 7, 1942 she became stranded near Capetown South Africa. She was refloated but sank while in tow.

Operating for American-Hawaiian Steamship Company as the Oklahoman, *the ex-*West Calera *is shown underway. The Mariners' Museum, Newport News, Va., neg. PB9233.*

West Camak

WPS Hull No.15
Tonnage:
    Gross   5647
    Net   3513
    Deadweight   7500
Length: 410.5 feet
Beam: 54 feet
Freeboard: 5 feet 9 1/2 inches
Draft: 24 feet 2 inches
Engine:
    Joshua Hendy Iron Works 3-cylinder
      triple expansion
    2800 IHP
    Oil fuel
Two decks
Crew 44.
Date Delivered: 1920. *Exact date and initial voyage not clear because of unreadable Movement Cards in the National Archives.*

From November 1920 until November 1921 she was in service between West Coast USA and European ports. The *West Camak* was tied up inactive at Camp Eustis, Hampton Roads from December 1, 1921 to February 18, 1922. From February 1922 to July 1923 *West Camak* shuttled between East Coast USA ports and South America. Following this she had a long period of service until May 1933 between Gulf Coast USA ports and European ports in Germany, Holland, Belgium, and England.

On June 10, 1933 at 11:30 AM *West Camak* was delivered to her purchaser Lykes Bros.-Ripley Steamship Company Inc. of New Orleans at Corpus Christi Texas with Corpus Christi becoming her port of registry. She remained in this service until 1938 when she was sold to Garibaldi S.A. Co-opertiua di Navigazione of Genoa, Italy. She was renamed *Antonio Locatelli* and registered at Genoa, Italy. The *Antonio Locatelli* continued in Italian service in the early World War II years. On November 12, 1940 she and three other freighters were in a small convoy passing between Albania and Southern Italy. This convoy was intercepted by a British cruiser and destroyer force under Vice Admiral Pridham-Wippell, and all were destroyed by gunfire after the escorts fled.

The West Camak *at anchor, probably awaiting a berth. Photograph Courtesy Peabody Essex Museum, neg. V-1374.*

West Camargo

WPS Hull No.16
Tonnage:

Gross	5881
Net	3659
Deadweight	7500

Length: 410.5 feet
Beam: 54 feet
Freeboard: 5 feet 9 1/2 inches
Draft: 24 feet 2 1/2 inches
Engine:
  Joshua Hendy Iron Works 3-cylinder
    triple expansion
  2800 IHP
  Oil fuel
Two decks
Crew 44.
Date Delivered: July 21, 1920.

Departed on maiden voyage on July 31 to New Zealand and Australia, returning November 29. She made a total of three trips to the South Pacific and was then transferred to Laidup Fleet Comm. on February 16, 1923. She was repaired in August 1923 and returned to active service in September 1923. From this time until June 1926 *West Camargo* remained in service between West Coast U.S.A. ports and South America.

On July 3, 1926 at 11:55 AM *West Camargo* was sold and delivered to Pacific Argentina Brazil Line at San Francisco with San Francisco being her port of registry. She remained in this service until ownership passed to Pope & Talbot Inc. of San Francisco in 1941. In 1942 the ownership reverted to the War Shipping Administration in Washington, D.C.

In 1943 *West Camargo* was transferred under lend-lease to the U.S.S.R. She was renamed *Desna* and stayed in Russian service until 1978. In the 1950's she was converted to a special cargo (fish) carrier with Nakhodka as her port of registry. This ship probably had the longest service of any WPS-built ship; some fifty-eight years.

Disposal: In 1978 *Desna* was sold to Japanese interests and was scrapped soon thereafter. She was renamed *Fuji Maru* but probable did not carry cargo for her Japanese owner.

The West Camargo *underway on a relatively calm sea. The Mariners' Museum, Newport News, Va., neg. PB15984.*

West Canon

WPS Hull No.17
Tonnage:

Gross	5645
Net	3517
Deadweight	8584

Length: 410.5 feet
Beam: 54 feet
Freeboard: 5 feet 9 1/2 inches
Draft: 24 feet 2 inches
Engine:

Joshua Hendy Iron Works 3-cylinder
   triple expansion
2800 IHP
Oil fuel
Two decks
Crew 43.
Date Delivered: September 1920.

The *West Canon* left on her maiden voyage to Cork, Ireland via Portland, Oregon on September 28, 1920, returning to San Francisco via Norfolk and Tiburon, California. There followed a voyage to the Far East with return through the Mediterranean Sea to New York. From January 1922 to October 1923 she remained in service between the East Coast U.S.A. and the Mediterranean and Far East points. In November 1923 the *West Canon* left Los Angeles for the Orient again returning to the East Coast via the Mediterranean Sea in April 1924. There followed one more trip to the Mediterranean and thence to India and Ceylon before returning to New York in November 1924. From then until July 1926 *West Canon* remained in service between East Coast U.S.A. ports and English ports.

On October 19, 1926 *West Canon* was delivered at Norfolk to the purchaser Ocean Transport Company of San Francisco. She was renamed *Pacific Spruce* and stayed in service of this company from 1926-1928. From 1928 to 1932 she was owned by Dimon Steamship Corporation. From 1932-1937 ownership reverted to the United States Shipping Board.

In 1937 *Pacific Spruce* was sold to Thompson Salmon Company of Seattle, Washington. She was renamed *William L. Thompson* and stayed in this company's service until February 1942. At that time she was delivered at Baltimore, Maryland to the U.S. Army on a bareboat charter. During February and March 1942 she was in the yard of Maryland Drydock Company at Baltimore undergoing ouerhaul and modification to a troopship. In June 1942 this ship began four years of service between the Northwest U.S.A. ports and Alaskan ports of Seward, Valdez, Ketchikan and Kodiak for the U.S. Army. In December 1942 a collision occured with the SS *Pacific Oak* requiring drydocking and repairs to her bow. From January to June 1945 extensive repairs were performed at Puget Sound Bridge and Dredge Corporation. On February 12, 1946 the Army returned the *Thompson* to the War Shipping Administration at Portland, Oregon. On the same day WSA delivered the *Thompson* to Pacific Atlantic Steamship Company and on May 6, 1946 she was delivered again to the Thompson Salmon Company.

Disposal: In 1947 the *William L. Thompson* was sold to China Merchants and renamed *Hai Fei*. On August 28, 1960 the *Hai Fei* arrived at Kaohsiung, Taiwan for scrapping which took place in September.

West Carmona

WPS Hull No.18
Tonnage:

Gross	5645
Net	3517
Deadweight	8584

Length: 410.5 feet
Beam: 54 feet
Freeboard: 5 feet 9 1/2 inches
Draft: 24 feet 2 inches
Engine:

    Joshua Hendy Iron Works 3-cylinder
      triple expansion
    2800 IHP
    Oil fuel
Two decks
Crew 44.
Date Delivered: October 1920.

Departed on maiden voyage to the Far East on November 5, 1920, returning on January 20, 1921. The *West Carmona* had two more trips to the Far East then was tied up in Southampton Bay in January 1922. Transferred to Laidup Fleet Comm. February 16, 1923. In July 1923 the *West Carmona* reentered active service between West Coast U.S.A. ports and the Far East. She remained in this service until March 1928.

On April 7, 1928 at 9:30 AM the *West Carmona* was delivered to the purchaser Oceanic and Oriental Navigation Company at San Francisco. She was renamed the *Golden State* with San Francisco as her port of registry.

In 1937 ownership changed to Matson Navigation Company of San Francisco and she was renamed *Lahaina*. On December 4, 1941 the *Lahaina* departed Ahukini, Territory of Hawaii, for San Francisco. On the afternoon of December 11 she was attacked by a Japanese submarine and shelled. The *Lahaina* was abandoned and then reboarded. She sank on December 12.

NOTE:
Four ships cancelled — WPS Hulls No.19-22. These were to be four Robert Dollar type steam freighters of 8800 deadweight ton capacity.

WPS Hulls Number 23 to 57. These were small craft such as barges built in the period between 1920 and 1940.

The West Carmona *alongside a pier with steam up and presumably ready to sail. San Diego Maritime Museum.*

American Manufacturer

WPS Hull No.58.
USMC Hull No. 94. Type C1-B.
Tonnage:

Gross	6778
Net	4800
Deadweight	8015

Length: 397 feet
Beam: 60 feet
Draft: 26 feet
Engines:
  2 Busch-Sulzer 7-cylinder diesel
Speed: 14 knots
Keel laid: February 5, 1940
Launched: August 8, 1940
Trial: March 24, 1941
Delivered: April 11, 1941.

The *American Manufacturer* was operated by U.S. Lines from 1941-1946. Thereafter she was operated by Agwilines Inc. of New York 1946-47. During this 1941-47 period ownership remained with the USMC. In 1949 the *American Manufacturer* was sold to Skibs. A/S Arizona of Oslo, Norway. The name was changed to *Hoegh Merchant* and the port of registry to Oslo, Norway. She remained in this service until 1956 when she was sold to Jacob Christensen of Bergen, Norway and renamed *Mathilda*. In 1964 she was sold to the Philippine President Lines Inc. and renamed *President Osmena*. Her port of registry became Manila and she was registered in the Philippines. In 1965 she was sold to the Seven Brothers Shipping Corp. of Manila and renamed *Seven Generals*.

Disposal: In 1973 the *Seven Generals* was scrapped in Kaohsiung Taiwan.

The American Manufacturer *underway as the* President Osmena. Photo: V.H. Young/L.A. Sawyer.

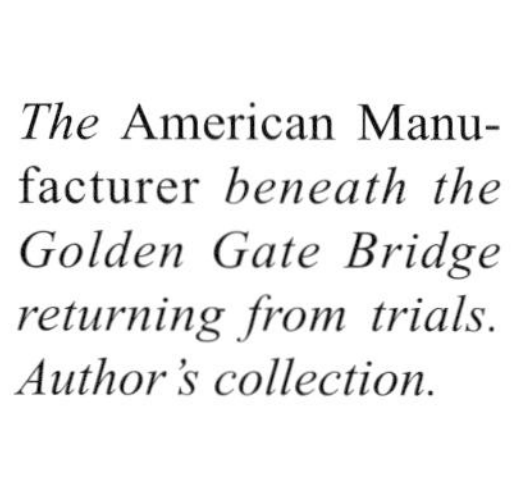

The American Manufacturer *beneath the Golden Gate Bridge returning from trials. Author's collection.*

American Leader

WPS Hull No. 58
USMC Hull No. 95 Type C1-B.
Tonnage:

Gross	6778
Net	4800
Deadweight	8015

Length: 397 feet
Beam: 60 feet
Draft: 26 feet
Engines:
    2 Busch-Sulzer 7-cylinder diesel
Speed: 14 knots
Keel laid: February 19, 1940
Launched: October 8, 1940
Trial: June 3 and 9, 1941
Delivered: June 12, 1941.

The *American Leader* was operated by U.S. Lines of New York for her brief life. On September 10, 1942 this ship was en route from Colombo to Newport News, Virginia and encountered the German surface raider *Michel* 800 miles west of the Cape of Good Hope. The *American Leader* was sunk by gunfire and the crew taken prisoner. Ten crew members were killed during the engagement.

Installation of the diesel engines, upper right and lower left, on the American Leader, *April 30, 1941. Western Pipe & Steel Co., photo no. 226, author's collection.*

American Builder

WPS Hull No.59.
USMC Hull No.96. Type C1-B.
Tonnage:

Gross	6778
Net	4800
Deadweight	8015

Length:	397 feet
Beam:	60 feet
Draft:	26 feet
Engines:	
2 Busch-Sulzer 7-cylinder diesel	
Speed:	14 knots
Keel laid:	August 15, 1940
Launched:	December 17, 1940
Trial:	July 21 and 22, 1941
Delivered:	July 25, 1941.

The *American Builder* was operated by U.S. Lines from 1941 to 1946 and thereafter by Agwilines Inc. of New York 1946-47. The ship remained under USMC ownership during her entire life.

In 1952 her name was changed to *Cape Decision* and she was placed in the James River Reserve Fleet. In 1968 *Cape Decision* was operated by the U.S. Army. The 1979 *Merchant Vessels of the United States* lists Cape Decision as "out of documentation."

The American Builder *at the outfitting dock, June 30, 1941. Stack painting has begun and the vessel is near completion. Western Pipe & Steel Co., photo no. 247, author's collection.*

American Press

WPS Hull No.60.
USMC Hull No.97. Type C1-B.
Tonnage:

Gross	6778
Net	4800
Deadweight	8015

Length:	397 feet
Beam:	60 feet
Draft:	26 feet
Engines:	
	2 Busch-Sulzer 7-cylinder diesel
Speed:	14 knots
Keel laid:	October 11, 1940
Launched:	March 11, 1941
Trial:	August 25 and 26, 1941
Delivered:	September 3, 1941.

The *American Press* was operated by U.S. Lines 1941 to 1945 and by Overseas Corporation of New York in 1946. Ownership remained with the USMC during her entire life. The name was changed to *Cape Lookout* in 1952.

Disposal: Sold by U.S. Department of Commerce to Hieruos Ardes in 1970. The *Cape Lookout* arrived in tow at Bilbao, Spain on December 21, 1970 for scrapping.

The American Press *under construction just after a brief shower, note wet deck. Western Pipe & Steel Co., photo no. 191, author's collection.*

American Packer

WPS Hull No.61.
USMC Hull No. 98. Type C1-B.
Tonnage:

Gross	6778
Net	4800
Deadweight	8015

Length:	397 feet
Beam:	60 feet
Draft:	26 feet
Engines:	
	2 Busch-Sulzer 7-cylinder diesel
Speed:	14 knots
Keel laid:	December 23, 1940
Launched:	May 20, 1941
Trials:	October 3 and 4, 1941
Delivered:	October 14, 1941

The *American Packer* was operated by U.S. Lines from 1941 to 1946 and again in 1947-48. In the 1946-47 period she was operated by A.H. Bull & Co. of New York. The *American Packer* remained under USMC ownership from 1941 to 1970.

Disposal: In 1970 she was sold to a Spanish company and taken to Valencia for scrapping.

The drama and excitement of a side launching are captured on May 20, 1941 as the American Packer *enters the water. Note the intensity of the onlookers as they watch the event. Western Pipe & Steel Co., photo no. 232, author's collection.*

Steel Artisan

WPS Hull No. 62
USMC Hull No. 171   Type C3-S-A2
Tonnage:
    Displacement: 7,800 tons
Length overall: 495 feet 8 inches
Beam: 111 feet 6 inches
Draft: 26 feet
Armament:
    2 5-inch antiaircraft
Speed: 18 knots
Complement: 860.
Keel laid: April 7, 1941.
Launched: September 27, 1941

The outfitting as a C-3 freighter was nearly complete when she was ordered converted to an escort aircraft carrier and re-named U.S.S. *Barnes* (AVG-7). The super-structure was removed and a flight deck was constructed. Trials were performed on August 12 and September 14, 1942. This ship was delivered to the British Navy on September 30, 1942 under the Lend-Lease aid program and was renamed HMS *Attacker*.

A British crew had previously arrived in San Francisco to man the ship. Also a flight of four Fairey Swordfish aircraft of 838 Squadron arrived at Alameda Naval Air Station from Nova Scotia in August 1942 to be embarked on HMS *Attacker*. She departed San Francisco December 11, 1942 and proceeded to Norfolk, Virginia via the Panama Canal. She left Curacao March 20, 1943 escorting a convoy of tankers and arrived on the River Clyde April 1. *Attacker* there underwent modifications which became standard for American built carriers operating in the Royal Navy. These included elongating the flight deck and changing the petrol storage and delivery system.

HMS *Attacker* initially joined the Western Approaches Command for convoy protection. On August 3, 1943 she departed for Gibraltar and the Mediterranean Fleet. With four other escort carriers *Attacker* participated in the Salerno landings operating Seafire fighters for troop support.

On September 20, 1943 she departed Bizerte, Tunisia and returned to England to the Western Approaches Command for convoy protection duty. In early 1944 *Attacker* was refitted in Liverpool for duty as an assault escort carrier. Following this she departed Belfast on May 14, 1944 with a destination of Gibraltar and the Mediterranean Fleet. On August 15, 1944 *Attacker* supported the landings in Southern France and in September was operating in the Eastern Mediterranean against Greek and Aegean Island targets. On November 3, 1944 HMS *Attacker* left Malta returning to Plymouth, England, but again returned to the Mediterranean where she underwent a three month refit at Taranto, Italy from December 1944 to March 1945.

After completion of the refit *Attacker* departed for the East Indies Station arriving Colombo on April 29, 1945. She spent the remaining war months in the Indian Ocean and was with forces accepting surrender of Japanese forces at Penang and Sabang in August 1945. HMS *Attacker* departed Bombay October 21, 1945 and arrived back on the Clyde November 11th.

On December 13, 1945 *Attacker* left Southampton for Norfolk, Virginia and return to the United States. On December 29th she was decommissioned from the Royal Navy and on September 13, 1946 the former *Attacker* was transferred to inactive status pending disposal.

In 1950 she was sold to Soc. Italiana Transporti Maritimi of Italy and converted to a passenger liner named *Castel Forte*.

In 1958 *Castel Forte* was sold to Fairline Shipping Co. of Monrovia and renamed *Fairsky*.

In 1978 *Fairsky* was again sold and converted to a floating hotel and restaurant. She was renamed *Philippine Tourist*.

Disposal: The last sale occurred in 1980 when she was sold to Li Chong Company Ltd. of Taiwan. *Philippine Tourist* arrived at Hong Kong on May 27, 1980 and dismantling commenced in June 1980 closing a long, glorious and varied career.

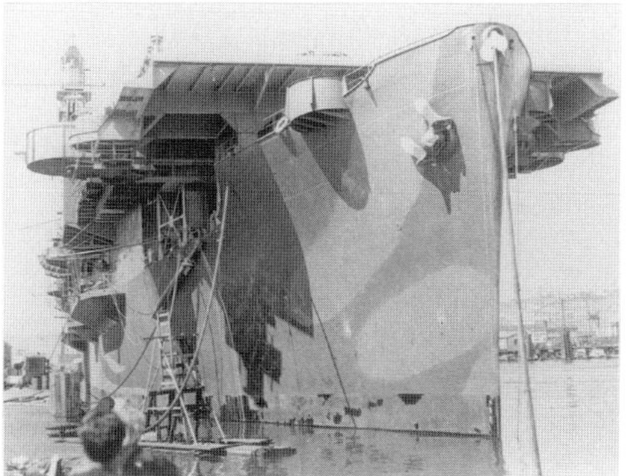

After almost being completed as the freighter Steel Artisan, *right, hull No. 62 had all its superstructure removed and was converted to the British escort carrier HMS* Attacker, *top right. After World War II the vessel underwent several name changes, including* Fairsky, *right center, and* Philippine Tourist, *right bottom. Right photo, Western Pipe & Steel Co.,photo no. 314, upper right, author's collection, bottom photos, courtesy of the Steamship Historical Society of America Collection, Langsdales Library, University of Baltimore.*

USS *Cascade* (AD-16)

WPS Hull No. 63
USMC Hull No. 172 Type C3-S1-N2
Tonnage:
    Displacement: 11,755 tons
Length overall: 492 feet
Beam: 69 feet 6 inches
Draft: 28 feet 6 inches
Armament:
    2 5-inch, 640 mm. antiaircraft
Machinery:
    Geared turbine
    9,350 SHP
Speed: 16.5 knots
Boilers: 2 Foster-Wheeler
Complement: 860.
Keel laid: July 17, 1941
Launched: June 7, 1942
Delivered: September 11, 1942

Cascade was removed from WPS when she was 70% complete because outfitting docks were full. She was taken to Matson Navigation Co. in San Francisco for completion and commissioned in the U.S. Navy March 12, 1943.

U.S.S. *Cascade* sailed from San Francisco on June 12, 1943 for Pearl Harbor where she was based servicing destroyers. In November she sailed westward and was stationed sucessively at Kwajalein, Eniwetok and Ulithi again servicing destroyers. From June until September 1945 *Cascade* was in Okinawa and then moved on to Wakayama Wan and Tokyo until March 1946.

U.S.S. *Cascade* was decommissioned to reserve at Philadelphia on February 12, 1947. She was recommissioned April 5, 1951 and stationed at Newport, Rhode Island. Subsequently there were training cruises to the Mediterranean and Caribbean Seas.

Disposal: *Cascade* remained in commission until 1975 when she was sold for scrapping to Luria Brothers of Brooklyn, New York. Dismantling took place at the Gulmar Yard in Brownsville, Texas starting September 1975.

USS Cascade *(AD-16), partially completed, left, looking from aft forward. As a destroyer tender, above, on station in the Pacific. Left photo; Western Pipe & Steel Co., photo no. 263, author's collection; upper, U.S. Navy.*

USS *Chandeleur* (AV-10)

WPS Hull No. 64
USMC Hull No. 173 Type C3-S1-B1
Tonnage:
    Displacement   14,250
Length overall: 492 feet
Beam: 69 feet 6 inches
Draft: 23 feet 9 inches
Armament:
    1 5-inch 38 caliber
    4 3-inch 50 caliber
    8 40mm. antiaircraft
Machinery:
    2 sets General Electric geared turbines
    9,350 SHP
Speed: 17.5 knots
Boilers: 2 Foster-Wheeler
Fuel Oil capacity:1,417 tons
Keel laid: May 29, 1941
Launched: November 29, 1941
Trial: November 9, 1942
Delivered and commissioned:
          November 19, 1942.

Chandeleur carried cargo from San Diego to Pearl Harbor, Efate, Espiritu Santo, Samoa and Noumea from January 15, 1943 to July 1943. *Chandeleur* was based at Espiritu Santo July 1 to November 13, 1943 as base for Patrol Squadron 71. She sailed between New Hebrides and Guadalcanal from October 1943 to March 1944 carrying men and aviation equipment. There then followed an overhaul on the West Coast. *Chandeleur* departed Oakland on May 18, 1944 for Kwalalein and Eniwetok. She supported Patrol Squadrons 202 and 216 at Eniwetok and Saipan through August 1944. In September she moved to Kossol Roads where she supported seaplanes for the Philippine campaign. She was tender for Patrol Squadron 21 at Ulithi from December 25, 1944 to February 8, 1945 when she moved to Kerama Retto to tend seaplanes for the Okinawa invasion. *Chandeleur* moved to Okinawa on July 15th and to Ominato Ko, Honshu after the Japanese surrender, leaving on October 16th for return to the West Coast and overhaul. In December 1945 *Chandeleur* made another trip to the Philippines to embark returning troops for return to Seattle. On February 12, 1947 U.S.S. *Chandeleur* was placed in reserve as an accommodations ship at Philadelpha. She continued on the list of U.S. Navy ships until 1970 and then was stricken.

USS Chandeleur *(AV-10), left, with keel and part of bottom plating in place, above as seaplane tender in the South Pacific. Left; Western Pipe & Steel Co., photo no. 285, author's collection; above, National Archives, 19-N-42823.*

USS *Hamlin* (AVG-15)

WPS Hull No. 65
USMC Hull No. 174 Type C3-S-A2
Tonnage:
    Displacement   13,700
Length overall: 492 feet
Beam: 69 feet 6 inches
Draft: 23 feet 3 inches, mean
                26 feet full load
Machinery:
    General Electric geared turbines.
    2 Foster-Wheeler boilers
    9,350 SHP
Speed: 18.5 knots
Fuel oil capacity: 3,100 tons
Aircraft: 20.
Armament: 2 4-inch antiaircraft
              14 20mm. antiaircraft
Complement: 646
Keel laid: October 6, 1941
Launched: March 5, 1942
Delivered: December 21, 1942

Delivered to British Navy after transfer under Lend-lease aid program. Name was changed to HMS *Stalker*. Departed for England via Panama and Norfolk, Virginia.

On reaching England she underwent the usual Royal Navy modifications and also was modified to become an assault support carrier. On September 9, 1943 *Stalker* supported the Salerno landings in company with HMS *Battler, Hunter* and *Attacker* using Seafire Mk III fighters. She departed for England escorting convoy MKF 24 and returned to the Mediterranean with convoy OS 77/KMS 51 in May 1944. From August 15-20, 1944 *Stalker* was one of a group of nine escort carriers supporting the landings in Southern France. *Stalker* participated in strikes in the Aegean Islands in September and October 1944. Following this *Stalker* passed through the Suez Canal and served in the Indian Ocean beginning in late 1944. She continued with service in the Eastern Fleet. On May 2, 1945 *Stalker* saw action in support of landings at Rangoon operating Seafire fighters. She participated in strikes against Burma and Sumatra in June and in mid-July against Cape Nicobar. Late August 1945 saw *Stalker* covering the reoccupation of Malaya and Singapore. HMS *Stalker* returned to England shortly thereafter and on December 2, 1945 she embarked troops at Southampton for return to Norfolk, Virginia. She was formally returned to the U.S. Navy at Norfolk Navy Yard on December 29, 1945. *Stalker* was decommissioned the same day and stricken from the U.S. Navy on March 20, 1946.

This ship was delivered to Waterman Steamship Co. at the Gulf Shipbuilding Corp., Mobile, Alabama, July 1, 1947 and converted to a standard C3-S-A2 freighter. She was subsequently sold to the Netherlands government on October 31, 1947 and renamed *Riouw*.

The *Riouw* was sold to Atlas Enterprises, Inc. of Panama in 1967 and renamed *Lobito*. The *Lobito* was sold again in 1971 to Comaran Africa Line of the Ivory Coast retaining the same name.

Disposal: In September 1975 *Lobito* was scrapped at Gandia, Spain.

NOTE:
WPS Hulls numbers 66 through 76 are not represented by large ships. The exact type of craft is not certain, but most probably were small barges.

Launched, above, as USS Hamlin *(AVG-15), the ship's name was changed to HMS* Stalker, *above right, after she was transferred to England under the Lend-Lease program. After the war she sailed as the* Riouw *under the Dutch flag. Above, Western Pipe & Steel Co., photo no. 348, author's collection; upper right, National Archives 19-5B4D-7; lower right, Skyfotos.*

USS *Croatan* (AVG-14)

WPS Hull No. 77
USMC Hull No. 191 Type C3-S-A2
Tonnage:
    Displacement   13,700 light
                            15,150 full load
Length overall: 492 feet
Beam: 69 feet 6 inches
Draft: 23 feet 3 inches, light
                   26 feet, full load
Machinery:
    General Electric geared turbines
    2 Foster-Wheeler boilers
    One shaft
    9,350 SHP
Speed: 18.5 knots
Fuel oil capacity: 3,100 tons
Aircraft: 20
Armament:
    2 4-inch
    14 20mm. antiaircraft
Complement: 646
Keel laid: September 5, 1941
Launched: April 4, 1942
Delivered: February 27, 1943

This ship was transferred to the British Navy under the Lend-Lease aid program and manned by a British crew in San Francisco on completion. She was renamed HMS *Fencer* and commissioned in the Royal Navy on March 1, 1943. *Fencer* departed San Francisco in March for New York, and thence to England with a load of 55 U.S. Army planes. On arrival in England *Fencer* was taken to Liverpool for modifications which included changing of aviation fuel storage and delivery, H/F direction finding gear, and radar. She was also prepared for operation in the Arctic.

In October 1943 HMS *Fencer* covered a convoy to the Azores to set up a RAF base. During this operation *Fencer*'s fighter planes operated from a new landing field in the Azores for thirteen days. From November 1943 to January 1944 *Fencer* escorted Gibraltar convoys and then protected Atlantic convoys in February and March. In early April 1944 *Fencer* was one of a group of carriers that attacked the German battleship *Tirpitz* in Norway. *Fencer*'s duty in this operation, named Tungsten, was to provide antisubmarine screening. April and May saw the *Fencer* protecting a convoy to Murmansk. She returned to Atlantic antisubmarine operations from June to October 1944. On November 1 HMS *Fencer* departed for Trincomalee (Ceylon) ferrying a load of aircraft. During 1945 she was used in similar capacity taking planes to Australia and Leyte for the British Far Eastern Fleet. She returned to England for a refit in October 1945 and at that time was adapted as a troop transport with bunking on the hanger deck. Between January and August 1946 there were several trips to Ceylon and Port Said as a troop transport. On October 22, 1946 *Fencer* departed Chatham enroute to Bermuda and Norfolk, Virginia. On November 21, 1946 she was officially returned to the U.S. Navy at Norfolk Navy Yard and was stricken from the U.S. Navy January 28, 1947.

The former *Fencer* was sold to National Bulk Carriers Inc. and delivered at Claremont, Virginia December 30, 1947. The plan to convert this ship to a tanker was found not feasible so she was converted to a C-3 cargo ship at Welding Shipyard of Norfolk, Virginia.

In 1950 this ship was sold to Achille Lauro of Italy and was taken to Naples for reconstruction as a passenger liner for the Italy to Australia route. She was renamed *Sydney* and in 1967 the name was again changed to *Roma*.

In 1971 the *Roma* was sold to Sovereign Cruise Ships of Cyprus and renamed *Galaxy Queen*. In 1972 she was again sold to G. Kotzouilis in Greece and renamed *Lady Dina*. In 1973 the name was changed to *Caribia 2*.

Disposal: In September 1975 *Caribia 2* was taken to Spezia, Italy for scrapping.

Laid down as USS Croatan *(AVG-14), upper right, the vessel became HMS* Fencer, *above, when transferred to the British under the Lend-Lease program. After the war, one of her many guises was that of the* Sydney, *lower right. Above photo, Imperial War Museum, lower right, photo: V.H. Young/L.A. Sawyer, upper right, Western Pipe & Steel Co., photo no. 323, author's collection.*

HMS *Striker*

WPS Hull No.78.
USMC Hull No.198. Type C3-S-A2.
Tonnage:

Displacement	13,700 light
	15,750 full load
Length:	492 feet
Beam:	69 feet 6 inches
Draft:	23 feet 3 inches, light
	26 feet, full load

Machinery:
General Electric geared turbines
Foster-Wheeler boilers
One shaft
9,350 SHP

Speed:	18.5 knots
Fuel oil capacity:	3,100 tons
Aircraft:	20

Armament:
2 4-inch
14 20mm. antiaircraft

Complement:	646.
Keel laid:	December 15, 1941
Launched:	May 7, 1942
Delivered:	April 28, 1943

Delivered to the British Navy under the Lend-Lease aid program. As originally planned this ship was to be named USS *Prince William* (AVG-19). She departed San Francisco in May 1943 and arrived at Liverpool on July 13th with convoy HH 246.

Striker underwent modifications for British service and completed trials in December 1943. She saw service in the Atlantic for antisubmarine protection from December 1943 through March 1944. On March 25th her aircraft accidentally shot down an American C54 thinking it to be a German FW 200. April 1944 was a busy month escorting convoy RA 59 to Russia and participating in an attack on the *Tirpitz*. In the May and June period operations against Norwegian targets and coastal shipping continued. In August strikes were made against shipping on the west coast of France. Later in August *Striker* returned to Arctic duty escorting convoys JW 59 and RA 59A while operating Composite Squadron 824. Her last European duty consisted of escorting convoy RA 60 to Russia in October 1944.

HMS *Striker* departed England in October, arriving Trincomalee, Ceylon November 22, 1944. She then shuttled between points in Australia, Manus, Leyte etc. acting as a replenishment carrier by carrying replacement planes for the fleet carriers of the British Pacific Fleet. In September 1945 *Striker* made a trip to transport prisoners of war from the Philippines to Australia. She returned to the Clyde River in December 1945, and subsequently departed Greenock on January 20, 1946 for the return to the United States. *Striker* was officially returned to the United States Navy at Norfolk, Virginia on February 12 and decommissioned the same day. *Striker* was stricken from the U.S. Navy on March 26, 1946.

Disposal: There was no reconversion of this ship and she was sold for scrap to Patapsco Scrap Corp. of Bethlehem, Pennsylvania on June 5, 1946.

HMS Striker *with bottom plating and part of double bottoms in place, right; just before launching, left; and moored to a buoy after several months of service. Right, Western Pipe & Steel Co., photo no. 325, left, Western Pipe & Steel Co., author's collection, lower, Imperial War Museum, neg. A21972.*

Sea Pike

WPS Hull No. 79
USMC Hull No. 267 Type C3-S-A2
Length overall: 492 feet
Beam: 69 feet 6 inches
Draft: 28 feet 6 inches
Keel laid: March 9, 1942
Launched: June 28, 1942
Delivered: February 13, 1943

Operated by Moore-McCormack Lines for the War Shipping Administration. The *Sea Pike* departed San Francisco on her first trip to the Pacific as a standard C-3 freighter. On return to San Francisco she underwent conversion to a troop carrier. As such she carried 2,109 troops and 235,170 cubic feet of cargo. From March 1943 to May 1945 the *Sea Pike* made repeated voyages to the Southwest and Western Pacific transporting troops and cargo to such places as Honolulu, Noumea, Australia, Philippines and Eniwetok. In June 1945 the *Sea Pike* departed San Francisco for La Havre and was in service between French ports and New York until December 1945.

On October 9, 1945 her name was changed to *Mormacwave*, since she had been sold to Moore-McCormack Lines of New York. As such she made one more trip to Marseilles and one trip to Japan and the Philippines before being released from troop carrying service.

Mormacwave was owned and operated by Moore-McCormack Lines from 1945-1955 when she was sold to States Marine Corporation of Delaware and renamed *Lone Star State*.

Disposal: She was sold for scrapping to a Taiwan firm in 1970 and was taken to Kaohsiung for dismantling.

At this stage of construction, May 1942, the Sea Pike looked more like a series of steel compartments than a ship. Note CVE at outfitting dock in background. Author's collection.

Sea Bass (I)

WPS Hull No. 80
USMC Hull No. 268 Type C3-S-A2
Length overall: 492 feet
Beam: 69 feet 6 inches
Draft: 28 feet 6 inches
Keel laid: April 10, 1942
Launched: August 2, 1942
Delivered: March 31, 1943

Operated by Matson Navigation Company for the War Shipping Administration. The *Sea Bass* departed San Francisco April 11 on her maiden voyage to Honolulu and subsequently made three trips to the South Pacific in 1943. On February 9, 1944 this ship was taken to Moore Drydock Company in Oakland where she was converted to a troop carrier, a process which took four months. *Sea Bass* had a capacity for 2,838 troops and 184,900 cubic feet of cargo.

The *Sea Bass* was in service between San Francisco and Port Hueneme, California and many ports in the South and Western Pacific from July 1944 to June 1945. On June 8, 1945 the ship left San Francisco on her only voyage to Europe going to LeHavre. She returned to New York and thence back to the Pacific in August. The *Sea Bass* continued, making three trips to the Philippines, Okinawa and Japan through March 1946. On March 26, 1946 she had an engine breakdown and required towing to Yokohama for repair. On arrival back in San Francisco June 3, 1946 the *Sea Bass* was released from troop carrying duty.

The *Sea Bass* was sold by USMC to the Luckenbach Steamship Company of New York in 1949 and was renamed the *William Luckenbach*. Registered in New York, she remained in service with Luckenbach from 1949-1960.

In 1961 the *William Luckenbach* was sold to Pope & Talbot, Inc. of San Francisco and was renamed the *P & T Forester*. This ownership lasted from 1961 to 1964.

In 1964 the *P & T Forester* was sold to American Foreign Steamship Company of New York. She was renamed the *American Robin* and remained in this service from 1964 to 1973.

Disposal: In October 1973 *American Robin* was sold to interests in Taiwan for scrapping.

The Sea Bass *at the outfitting dock on March 2, 1943, has some of her armament in place and is nearing completion. Author's collection.*

Sea Angel

WPS Hull No. 81
USMC Hull No. 269 Type C3-S-A2
Keel laid: May 13, 1942
Launched: September 7, 1942
Renamed USS *Boliver* (APA-34) during construction.
Delivered: March 15, 1943

Commissioned in the U.S. Navy the day she was delivered, the ship departed for New York where she was decommissioned April 23 at Hoboken, New Jersey for conversion to an attack transport. Recommissioning took place September 1, 1943. *Bolivar* left New York in October, returning to the West Coast for landing training. She arrived Lahaina Roads, Territory of Hawaii January 21, 1944. *Bolivar* shuttled troops over the Pacific and participated in invasions of Kwajalein, Saipan, Guam, Yap, Leyte, Lingayan Gulf, and Iwo Jima. In June 1945 she returned to Portland for an overhaul and did not participate in another invasion. *Bolivar* spent the time from September 1945 through January 1946 returning troops from the Philippine, Marshall, Admiralty and Caroline Islands. Following this she departed for Norfolk, Virginia, thence to the New York Navy Yard where she was decommissioned April 29, 1946 and returned to the Maritime Commission September 12, 1946. The USS *Bolivar* received five battle stars for service in the Pacific.

In March 1949 the *Bolivar* was sold to American President Lines and was renamed *President Van Buren*, being the third ship so named. She carried cargo and twelve passengers. On October 8, 1967 her name was changed to *President Harding*.

On March 25, 1968 this ship was sold to the Pacific Far East Lines and renamed *Thailand Bear*. In 1970 she was sold to Grace Prudential and renamed *Santa Monica*.

Disposal: The *Santa Monica* was scrapped in Kaohsiung, Taiwan in February 1972.

The Sea Angel, *right, as USS* Bolivar, *in San Francisco Bay returning veterans from the Pacific, and, above, as the* President Van Buren *(III) at a pier in San Francisco Bay. Right, author's collection, above, APL Archives.*

Sea Mink

WPS Hull No. 82
USMC Hull No. 270 Type C3-S-A2
Length overall: 492 feet
Beam: 69 feet 6 inches
Draft: 28 feet 6 inches
Keel laid: June 10, 1942
Launched: October 10, 1942
Renamed USS *Callaway* (APA-35) during construction.
Delivered: April 24, 1943

Commissioned in the U.S. Navy the day she was delivered, *Callaway* departed San Francisco and was on the East Coast until departing Norfolk October 23, 1943. She went to Hawaii via San Diego and during the next eighteen months was occupied shuttling troops around the Pacific Ocean. *Callaway* participated in the landings at Kwajalein, Emirau, Saipan, Pelaus, Leyte, Lingayan Gulf, and Iwo Jima. On January 8, 1945 *Callaway* was struck on the bridge by a kamikaze plane causing minimal damage but killing twenty-nine men and wounding twenty-two. Temporary repairs were made at Ulithi and *Callaway* resumed service until returning to San Francisco in June for overhaul. She arrived back in Hawaii August 27, 1945, and then sailed to Wakayama, Japan with occupation troops. *Callaway* made two trips from the Far East to the West Coast returning veterans and then sailed to New York where she was decommissioned on May 10, 1946.

The former *Callaway* was sold to American President Lines in March 1949 and was renamed *President Harrison*. In service she carried cargo and 12 passengers. On March 10, 1966 her name was changed to *President Fillmore* (IV).

On April 24, 1968 she was sold to the Waterman Steamship Company and renamed *Hurricane*.

Disposal: The *Hurricane* was scrapped at Kaohsuing, Taiwan in 1974.

USS Callaway, *a veteran of the Marshalls, Marianas, Palaus and Iwo Jima campaigns, shown at anchor during World War II. U.S. Coast Guard.*

The Sea Mink *as the* President Harrison *at a pier in the San Francisco Bay area. APL Archives.*

Sea Swallow

WPS Hull No. 83
USMC Hull No. 271 Type C3-S-A2
Tonnage:
    Displacement: 16,100
Length overall: 492 feet
Beam: 62 feet
Maximum Draft: 26 feet 6 inches
Armament:
    2 5-inch
    4 40mm. antiaircraft
Machinery:
    Geared turbines
    2 Combustion Engineering boilers
    9,350 SHP
Speed: 18 knots
Keel laid: July 1, 1942
Launched: November 10, 1942
Renamed USS *Cambria* (APA-36) during construction.
Delivered: May 4, 1943.

Cambria departed San Francisco the day she was delivered for New York where she was decommissioned and converted to an attack transport. Recommissioning took place on November 10, 1943. *Cambria* left New York December 11 bound for Pearl Harbor via Norfolk. She departed Pearl Harbor January 23, 1943 and acted as flagship for the landings on Majuro. Following this she returned to San Francisco for an overhaul before returning to Pearl Harbor. *Cambria* participated in the Saipan landings on June 15, 1944 and then was flagship for the Tinian landings July 24 to August 1, 1944. This ship was involved in the Leyte landings in November 1944 and the Linguayan Gulf landings on January 10, 1945. *Cambria* returned to Tulagi and thence Ulithi before participating in the Okinawa landings April 1, 1945. She departed Okinawa April 3 for San Pedro and an overhaul lasting until August. The war had ended by the time of her return and she spent the remainder of 1945 returning troops from the Far East.

U.S.S. *Cambria* departed San Francisco January 11, 1946 for Norfolk, Virginia and duty in the Atlantic. She was placed in reserve June 30, 1949, but was again recommissioned September 15, 1950, and participated in landing exercises in the Caribbean and Labrador. In 1956 *Cambria* landed United Nation troops at Gaza during the Suez Crisis. Returning to Norfolk on February 2, 1957, she was involved in training exercises for landings. In September 1958 *Cambria* supported landings at Beirut, Lebanon. She returned to the United States in March 1959 and visited Great Lakes ports in June and July in conjunction with the opening of the St. Lawrence Waterway.

The U.S.S. *Cambria* received six battle stars for World War II service. She was stricken from the U.S. Navy September 14, 1970. She was sold by the Navy to Haitian interests in February 1972. The plan was to use *Cambria*, renamed *Jean Claude Duvalier*, as a floating warehouse at Tortuga. Financial difficulties ensued and *Cambria* never left Norfolk.

Disposal: In February 1972 She was sold for scrapping at Portsmouth, Virginia.

USS Cambria *at anchor on March 26, 1952. As an attack transport she carried troops, landing craft and sufficient armament to fight off attacking aircraft. U.S. Navy.*

Sea Snipe

WPS Hull No. 84
USMC Hull No. 272 Type C3-S-A2
Length overall: 492 feet
Beam: 69 feet 6 inches
Draft: 28 feet 6 inches
Keel Laid: August 8, 1942
Launched: December 7, 1942.
Delivered: May 29, 1943

Operated by American President Lines for the USMC, the *Sea Snipe* carried 225,500 cubic feet of cargo and 2,194 troops. The ship departed San Francisco for Australia August 13, 1943. She spent the time until July 1945 shuttling between the West Coast and points in the South and Western Pacific. Between July and December 1945 the *Sea Snipe* carried troops from France and Italy back to the East Coast of the USA. In December 1945 she developed serious boiler trouble requiring a major overhaul and cancelling further voyages to Europe.

In 1946 the *Sea Snipe* was operated by Dickman, Wright & Pugh. In 1947 the USMC sold her to Luckenbach Steamship Company of New York and she was renamed *Edward Luckenbach* with her port of registry being New York. She remained in service with this line from 1947 to 1959.

The *Edward Luckenbach* was sold to States Marine Lines Inc. of New York in 1959. Her name was changed to *Aloha State*. Service continued with States Marine Lines from 1959 to 1971.

Disposal: In 1971 the *Aloha State* was sold to interests in Taiwan where she was taken for scrapping.

The Sea Snipe *as she appeared in December, 1952 as the* Edward Luckenbach. *Photograph Courtesy Peabody Essex Museum, neg. 23,725.*

Sea Needle

WPS Hull No. 85
USMC Hull No. 273. Type C3-S-A2
Tonnage:
    Displacement   16,100
Length overall: 492 feet
Beam: 69 1/2 feet
Maximum draft: 26 1/2 feet
Armament:
    2 5-inch
    4 40mm. antiaircraft
Machinery:
    Geared turbines
    2 Combustion Engineering boilers
    9,350 SHP
Speed: 18 knots
Keel laid: September 10, 1942
Launched: December 29, 1942
Renamed U.S.S. *Chilton* (APA-38) during construction.
Delivered: May 29, 1943.

She departed San Francisco for the New York Navy Yard where conversion to an attack transport took place. *Chilton* was commissioned December 7, 1943 and became a training ship for attack transport crews at Newport from January to October 1944. She then departed for the Pacific, arriving at Pearl Harbor January 23, 1945. *Chilton* landed troops at Kerama Retto, Okinawa March 26, 1945 and continued on at Okinawa as the flagship of Transport Squadron 17. An overhaul at San Francisco then followed, but she was back at Ulithi in July. With the advent of peace *Chilton* was involved in redeployment of troops in the Far East and returning others to the USA.

Chilton left San Francisco for the Bikini Atomic Tests on June 2, 1946 and returned August 1. From September 1946 to January 1947 there followed a tour of duty in the Far East. She was a floating laboratory at Bikini in the summer of 1947. *Chilton* stayed in active service until 1963 and during this time saw service in China, the Caribbean, and Mediterranean theaters. This ship received one battle star for her World War II service.

Disposal: U.S.S. *Chilton* was stricken from the U.S. Navy July 1, 1972 and was sold for scrapping at Camden, New Jersey. It is to be noted that this ship was the last of over 388 APA's and AKA's constructed during World War II to remain in Navy service.

At the outfitting dock the Sea Needle *has yet to have her name changed to USS* Chilton *and be modified to an APA. Author's collection.*

Sea Carp

WPS Hull No. 86.
USMC Hull No. 274. Type C3-S-A2
Keel laid: October 14, 1942
Launched: January 23, 1943
Renamed USS *Clay* (APA-39) during construction.
Delivered: June 29, 1943.

She proceeded to New York where she was converted to an attack transport between August and December 1943. The *Clay* arrived at Pearl Harbor in February 1944 and sailed over 188,888 miles in Pacific operations before the war ended in August 1945. She participated in landings at Saipan, Leyte, Lingayan Gulf and Okinawa. *Clay* was flagship Transport Division 18 for the Saipan operations and flagship Transport Division 13 in September 1945. In October *Clay* carried occupation troops to Tientsin. The remainder of 1945 was occupied with the return of troops from the Far East.

The *Clay* sailed from San Francisco March 9, 1946 for New York where she was decommissioned May 15th.

On September 12, 1946 the former U.S.S. *Clay* was sold by the USMC to American President Lines. She was renamed *President Johnson* (II) and remained in APL service until 1968. On April 24, 1968 she was sold to Waterman Steamship Company and renamed *La Salle*.

Disposal: In November 1974 the *La Salle* was sold to the Zui Feng Steel Company and scrapped at Kaohsiung, Taiwan.

The Sea Carp *after her name was changed to* President Johnson *(II) in 1949 underway on a light sea dotted with whitecaps. APL Archives.*

USS *Bayfield* (APA-33)

WPS Hull No. 87
USMC Hull No. 275 C3-S-A2
Formerly named *Sea Bass* (II).
Tonnage:
　　Displacement:　　　　16,100
Length overall:　492 feet
Beam:　　　　　69 1/2 feet
Maximum draft:　26 1/2 feet
Armament:
　　2 5-inch
　　4 40mm. antiaircraft
Machinery:
　　geared turbines
　　2 Combustion Engineering boilers
　　9,350 SHP
Speed:　　　　　18 knots
Keel laid:　　　November 11, 1942
Launched:　　　February 15, 1943
Delivered:　　　June 30, 1943.

Departed San Francisco July 7 for New York where she was converted to an attack transport. *Bayfield* was recommissioned in New York on November 30, 1943. The ship participated in training exercises on the East Coast and then was involved in the practice landings for Normandy conducted at Slapton Sands. The surprise attack by German ships resulted in sinking of LST's but *Bayfield* was not damaged. At the Normandy landings on June 6 *Bayfield* was flagship of CTF 125 at Utah Beach. In July and August she was in the Mediterranean and participated in the landings in Southern France as flagship of CTF 87. In September *Bayfield* returned to Norfolk for repairs and thence to the Pacific. In February 1945 she was involved in the Iwo Jima landings and in April the Okinawa landings. *Bayfield* then carried cargo to Pacific bases until her return to San Francisco in July. From August 1945 to March 1946 she shuttled troops and cargo to Japan and Korea. From April to August 1947 *Bayfield* participated in the Bikini atomic tests. There were two further voyages to China between November 1948 and March 1949.

In May 1949 *Bayfield* was transferred to Atlantic Fleet duty, but was back in the Pacific for the Korean War. She participated in the Inchon landing in September 1950 and the Wonson landing in October. In December 1950 she took part in the Hungnan evacuation. Between 1951 and 1954 there were several trips to ports in Japan, Korea and Vietnam.

U.S.S. *Bayfield* received four battle stars for World War II service and four battle stars for Korean War service. She was stricken from the U.S. Navy on October 10, 1968 and returned to the USMC.

Disposal: In September 1969 *Bayfield* was sold for scrapping to San Jose, California shipbreakers.

USS Bayfield (APA-33) *brings soldiers home shortly after World War II. Painted on the bluff off her bow is "Welcome Home, Well Done." Author's collection.*

USS *Cavalier* (APA-37)

WPS Hull No. 88
USMC Hull No. 276 Type C3-S-A2
Tonnage:
    Displacement   16,100
Length overall: 492 feet
Beam: 69 1/2 feet
Maximum draft: 26 1/2 feet
Armament:
    2 5-inch
    4 40mm. antiaircraft
Machinery:
    geared turbines
    2 Combustion Engineering boilers
    9,350 SHP
Speed: 18 knots
Keel laid: December 10, 1942
Launched: March 15, 1943
Delivered: July 19, 1943

Departed San Francisco on the day of delivery for Hoboken, New Jersey where she was converted to an attack transport by Bethlehem Steel Company. She was commissioned in the U.S. Navy January 15, 1944. In January *Cavalier* departed for the Pacific with two construction batallions on board.

She participated in the Saipan landings in June and Tinian landings in July 1944. *Cavalier* was involved also in the Leyte landings October 1944 and the Lingayan landings in January 1945. At this time she had casualties from kamikaze shrapnel. *Cavalier* landed troops in Manila Bay on January 30, 1945. She was torpedoed by Japanese submarine *RO 115* and had 50 men injured. *Cavalier* was towed to Leyte for temporary repairs and then proceeded to Pearl Harbor for permanent repair lasting until September. With the war over she carried troops between Western Pacific ports and the USA. She had two tours of duty in China between 1946 and 1948. During the Korean War *Cavalier* was again involved and participated in landings at Pohang, Inchon and Wonson.

U.S.S. *Cavalier* received five battle stars for World War II service and four battle stars for Korean War service. She was stricken from the U.S. Navy on October 1, 1968 and returned to the USMC.

Disposal: *Cavalier* was placed in the Marad Reserve Fleet October 1966 and was scrapped in Hong Kong August 1969.

USS Cavalier *(APA-37) entering San Francisco Bay on November 1, 1945. Note the skyline in the background. Painted under the landing craft amidships is "The 716 Tank Battalion, The Wolfpack Returns," followed by a listing of the battles they were in. Author's collection.*

Sea Barb

WPS Hull No. 89.
USMC Hull No. 277 Type C3-S-A2
Keel laid: December 31, 1942
Launched: April 8, 1943
Delivered: August 6, 1943

The ship was chartered and operated by the U.S. Army on August 6, 1943. The *Sea Barb* underwent conversion to a troopship by Marine Repair Shop of San Francisco Port of Embarkation and was completed December 1943. She had a capacity of 187,900 cubic feet of cargo and 2,911 troops.

The *Sea Barb* departed San Francisco for the Southwest Pacific in January 1945. She spent the time until June of 1946 shuttling between ports in the Far East and the South Pacific and San Francisco. She returned to the Fort Mason Marine Repair Shop for major repairs in mid 1944. After June of 1946 *Sea Barb* was released from U.S. Army service and returned to the USMC.

In 1949 the *Sea Barb* was sold by the USMC to Luckenbach Steamship Company of New York. She was renamed *F.J. Luckenbach* and was homeported in New York. She remained in service of Luckenbach from 1949 to 1961 when she was sold to Pope & Talbot Inc. of San Francisco. Renamed the *P & T Seafarer* she was registered in San Francisco and remained with Pope & Talbot from 1961 to 1964.

In 1964 this ship was sold to American Foreign Steamship Corporation of New York. She was renamed *American Hawk* and registred in New York. Service with American Foreign Steamship continued from 1964 to 1971. On June 14, 1971 an underwater explosion occured when the *American Hawk* was in Qui Nhon, Vietnam at a military dock. She subsequenly settled to the bottom but was later refloated. Taken to Hong Kong for docking and examination, leakage continued and it was elected to scrap the ship locally in Hong Kong in September 1971.

The Sea Barb *as the* American Hawk *alongside a San Francisco pier on July 13, 1965. Author's collection.*

Sea Cat

WPS Hull No. 90
USMC Hull No. 278 Type C3-S-A2
Keel laid: January 26, 1943
Launched: May 4, 1943
Delivered: August 25, 1943

Chartered by the War Department the day she was delivered, the *Sea Cat* was operated by the U.S. Army. She departed San Francisco for Honolulu as a freighter in September 1943. On return to San Francisco she was taken to Marine Repair Shop of San Francisco and underwent conversion to a troopship between October 1943 and February 1944. The *Sea Cat* had a capacity of 191,000 cubic feet of cargo and 1,979 troops. Between February 1944 and June 1945 she shuttled between San Francisco and many ports in the South and Western Pacific. Between July and December 1945 there were three round trips to Europe returning troops from Italy and France. The *Sea Cat* was again in the Pacific from January until April 1946, at which time she spent two months overhauling.

The *Sea Cat* was returned to the USMC in 1946. She was then sold to Luckenbach Steamship Company of New York in 1949 and renamed *Lena Luckenbach* and registered in New York. The ship remained in service of this company from 1949 until 1964 when she was transferred to Luckenbach Overseas Corporation from 1964 to 1968. She was renamed *Overseas Lena* in 1968 with homeport remaining New York. In 1969 the *Overseas Lena* was transferred to Overseas Carriers Corporation and her name was again changed, this time to *Overseas Eva*. This service continuing until 1971.

Disposal: In 1971 the *Overseas Eva* was sold to interests in Taiwan, and was scrapped in Kaohsiung, Taiwan.

Loaded with troops and with flags flying and steam up, the Sea Cat *is about to embark. Note the debarkation nets rolled up at the edge of the deck at each hold. Steamship Historical Society of America Collection, Langsdale Library, University of Baltimore.*

Sea Devil

WPS Hull No. 91
USMC Hull No. 279. Type C3-S-A2
Keel laid: February 18, 1943
Launched: May 23, 1943
Delivered: November 30, 1943

Entering service December 8, 1943 under operation of American Hawaiian Steamship Company for the War Shipping Administration, the *Sea Devil* carried 250,000 cubic feet of cargo and 2,101 troops. Between December 1943 and January 1946 this ship plied between San Francisco and many ports in the South and Western Pacific. In February *Sea Devil* picked up a load of 2,040 prisoners of war in Seattle and returned them to Europe. She continued in service between New York and Le Havre until June 1946 when troop carrying duty ceased.

In 1949 the *Sea Devil* was sold to Luckenbach Steamship Company. She was renamed *Harry Luckenbach* and registered in New York. In 1959 the *Harry Luckenbach* was sold to States Marine Lines of New York. The name was changed to *Copper State* and she remained with this company from 1959 to 1973.

Disposal: In April 1973 *Copper State* was scrapped in Kaohsiung, Taiwan.

As the Harry Luckenbach *the ex-Sea Devil* *is shown alongside a pier on September 20, 1950. Note the raft and rope ladder alongside no. 3 hatch and the staging and rope ladder at no. 1 hatch, indicating the crew is probably painting the offshore side and has gone to coffee. Photograph Courtesy Peabody Essex Museum, neg. 23,729.*

Sea Flasher

WPS Hull No. 92
USMC Hull No. 280 Type C3-S-A2
Keel laid: March 18,1943
Launched: June 22, 1943
Delivered: December 24, 1943

Beginning service on January 10, 1944 with Isthmian Steamship Company, the ship was operated for the War Shipping Administration. She functioned as a troopship carrying 226,500 cubic feet of cargo and 2,086 troops. From January 1944 until March 1946 the *Sea Flasher* operated between the West Coast ports of San Francisco, Portland and Seattle and many ports in the Western Pacific.

In February 1946 the *Sea Flasher* was released from troop service and sent to Hampton Roads being assigned to the WSA Reserve Fleet at Lee Hall, Virginia on April 17, 1946.

In 1947 the USMC sold the *Sea Flasher* to the Isthmian Steamship Company of New York. She was renamed *Steel Age* and New York remained her port of registry during this ownership which lasted from 1947 to 1971.

Disposal: The *Steel Age* was sold to interests in Taiwan in 1971 and was scrapped in April 1971 at Kaohsiung.

Shown at the outfitting dock, the Sea Flasher *nears completion. Note the yet to be armed antiaircraft gun tubs and uninstalled ventilators on the flying bridge. Western Pipe & Steel Co., photo no. 753, author's collection.*

Sea Arrow

WPS Hull No. 93
USMC Hull No.281 Type C3-S-A2
Keel laid:　　　April 12, 1943
Launched:　　July 10, 1943
Renamed USS *Alpine* (APA-92) during construction.
Delivered:　　September 30, 1943.

Commissioned in the U.S. Navy April 22, 1944, the *Alpine* took part in the Guam landings in July and the Leyte landings in October. In the Leyte operation she was hit by a kamikaze killing three and retired to Manus for repair. The U.S.S. *Alpine* participated in the Lingayan Gulf landings in January 1945 and later the same month supported the landings in Subic Bay. In the March-April period she acted as a repair ship for landing craft. The *Alpine* was involved in the Okinawa landings in March and was hit again by a kamikaze killing thirty-three. Temporary repairs were performed at Okinawa and she returned to Seattle for permanent repair. The *Alpine* was back in Korea in October-November and made a final trip in December

of 1945. After return from this trip *Alpine* was sent to Norfolk in February 1946 and decommissioned April 5, 1946. This ship was transferred back to the USMC April 10, 1946. She received five battle stars and a U.S. Navy unit citation for service in World War II.

In 1947 the ex-*Alpine* was sold by the USMC to American Mail Line Ltd. of Seattle. The name was changed to *India Mail* and Portland, Oregon became her port of registry. She continued in American Mail Line service from 1947 to 1965.

In 1965 the *India Mail* was sold to Hudson Waterways Corporation of New York. Her name was changed to *Transwestern* and registration to New York during this ownership which lasted until 1969.

The Buckeye Steamship Company of Delaware purchased this ship in 1969, renaming the vessel *Buckeye Pacific*. This ownership continued from 1969 to 1971.

In 1971 another sale took place, this time to Empire Steamship Company of Panama and the ship was renamed *Empire Pacific*.

Disposal: In 1971 *Empire Pacific* was sold to Taiwanese interests for scrapping. She arrived at Kaohsiung on December 7,1971.

Near to completion, the Sea Arrow *is shown at the outfitting dock. Although her kingposts are in place, the masts and cargo gear have not yet been installed, nor have her guns. Western Pipe & Steel Co., photo no. 731, author's collection.*

Sea Snapper

WPS Hull No. 94
USMC Hull No. 282 Type C3-S-A2
Keel laid: May 6, 1943
Launched: August 5, 1943
Renamed USS *Barnstable* (APA-93) during construction.
Delivered: October 30, 1943

Commissioned in the U.S. Navy the day of delivery, she was taken to the Commercial Iron Works in Portland, Oregon where conversion to an attack transport took place. *Barnstable* arrived Pearl Harbor in July 1944 to join Transport Division 32. After practice exercises at Guadalcanal she was in on the Peleliu landings in September 1944 and the Ulithi occupation later the same month. The *Barnstable* participated in both the Leyte landings in October and the Lingayan Gulf landings in January 1945. There followed several trips to shuttle more troops into the Philippines, and then she took part in the Okinawa invasion in April. *Barnstable* returned to the West Coast for overhaul in April and then returned to the Far East where she shuttled occupation troops until January 1946. *Barnstable* arrived on the East Coast in February where she was inactivated and decommissioned March 25, 1946. Return to the USMC followed.

The U.S.S. *Barnstable* received four battle stars for World War II service.

In 1948 the USMC sold the ex-*Barnstable* to Isthmian Steamship Company of New York. The ship was renamed *Steel Fabricator* and she stayed in service with this company from 1948 to 1969.

The *Steel Fabricator* was sold to Reliance Carriers SA of Panama in 1971 and was renamed *Reliance Dynasty*. In 1972 she was again sold, this time to Valor Navigation Corporation of Panama and renamed *Grand Valor* for her service with this company which lasted until 1973.

Disposal: In March 1973 the *Grand Valor* was scrapped in Kaohsiung, Taiwan.

The Sea Snapper *on the launching way. Note the crane operator, lower right, watching camera man hanging from a crane "bucket" to shoot the above photo. Western Pipe & Steel Co., photo no. 733, author's collection.*

Sea Corporal

WPS Hull No. 95
USMC Hull No. 283 Type C3-S-A2
Keel laid: May 25, 1943
Launched: August 26, 1943
Delivered: January 31, 1944

The *Sea Corporal* was operated by American President Lines for the War Shipping Administration. She was equipped as a troop transport with a capacity of 2,165 troops and 227,536 cubic feet of cargo.

The *Sea Corporal* departed San Francisco February 7, 1944 for Brisbane, Australia, on her maiden voyage. She stayed in service between the West Coast and many ports in the South and Western Pacific until December 1945. *Sea Corporal* departed Los Angeles February 16, 1946 with 2,000 prisoners of war to be returned to Le Havre. On return from France in March she was released from troop service and laid up in the WSA Reserve Fleet at Lee Hall, Virginia April 25, 1946.

The USMC sold the *Sea Corporal* to Isthmian Steamship Company of New York in 1947 and she was renamed *Steel Surveyor*. Registered in New York she remained in service of this company from 1947 to 1971.

Disposal: In 1971 the *Steel Surveyor* was sold to Taiwanese interests and scrapped in July of 1971.

NOTE:
WPS Hulls No. 96 through 120 were destroyer escorts contracted to the WPS San Pedro yard. Hulls 96 through 107 were completed, the remainder were cancelled.

*The ex-*Sea Corporal *under way as the* Steel Surveyor *for the Isthmian Steamship Company. The condition of her hull would indicate many years of hard use. Steamship Historical Society of America Collection, Langsdale Library, University of Baltimore, no. 8886.*

Sea Angler

WPS Hull No. 121
USMC Hull No. 1544 Type C3-S-A2
Keel laid: June 24, 1943
Launched: September 27, 1943
Renamed USS *Cecil* (APA-96) during construction.
Delivered: February 26, 1944

Placed in reduced commission in the U.S. Navy February 27, *Cecil* was taken to the Commercial Iron Works in Portland, Oregon for conversion to an attack transport. U.S.S. *Cecil* was placed in full commission on September 15, 1944 after completion of the conversion. She departed San Francisco November 26th for the Western Pacific where she shuttled troops and cargo until January 1946. During this time she participated in the landings at Iwo Jima and Okinawa. The *Cecil* was transferred to Norfolk, Virginia in March where she was decommissioned and returned to the USMC.

The U.S.S. *Cecil* received two battle stars for service in World War II.

In 1948 the Isthmian Steamship Company purchased the ex-*Cecil* from the USMC and renamed her *Steel Admiral* and registered her in New York.

Disposal: In 1973 the *Steel Admiral* was sold to Taiwanese interests and scrapped in October 1973 at Kaohsiung.

The Sea Angler *quickly became the USS* Cecil *(APA-96), above. Launched late in the war, she shows a dazzle-painted hull, a rare practice so late in the war. Photograph Courtesy Peabody Essex Museum, neg. 23,727.*

Sea Fiddler

WPS Hull No. 122
USMC Hull No. 1545 Type C3-S-A2
Keel laid: July 13, 1943
Launched: October 16, 1943
Delivered: May 18, 1944.

Completed as a troop transport, she carried 2,107 troops and 258,000 cubic feet of cargo. The *Sea Fiddler* was operated by Isthmian Steamship Company for the WSA.

This ship departed San Francisco June 1, 1944 on her maiden voyage to the Southwest Pacific and Guam. She continued service to the Pacific theater until June 1945 when she departed San Francisco for Leghorn and Naples, Italy. Until December 1945 the *Sea Fiddler* stayed in service between the East Coast and points in Europe and North Africa. In January 1946 there was one voyage from Hampton Roads to Yokohama with return to Los Angeles. Two more trips to Le Havre followed returning prisoners of war to Europe. On return to New York in May 1946 the *Sea Fiddler* was released from troop carrying service.

On May 14, 1947 the *Sea Fiddler* was sold to the Matson Navigation Company of San Francisco. She was renamed *Hawaiian Refiner* and carried twelve passengers and 4,945 tons of cargo in service between California and Hawaii. This service continued from 1947 to 1971.

Disposal: The *Hawaiian Refiner* was sold to American Ship Dismantlers of Los Angeles on April 12, 1971. Subsequently she was sold to Tung Ho Steel Company of Kaohsiung, Taiwan and taken there under tow May 22, 1971. Dismantling started in Kaohsiung August 5, 1971.

Launched as the Sea Fiddler, *hull 122 later became the* Hawaiian Refiner *for Matson Navigation Co. Western Pipe & Steel Co., photo no. 795, author's collection.*

*Underway on a glassy sea, the ex-*Sea Fiddler *as the* Hawaiian Refiner *for Matson Navigation Co. Matson Lines photo.*

Sea Flier

WPS Hull No. 123
USMC Hull No. 1546 Type C3-S-A2

Keel laid: August 7, 1943
Launched: November 16, 1943
Delivered: May 27, 1944.

The *Sea Flier* was completed as a troop transport carrying 2,027 troops and 242,136 cubic feet of cargo. She was operated by Moore-McCormack Line for the WSA.

The *Sea Flier* operated between the West Coast and far Pacific points from June 1944 until June 1945. At that time she sailed to Marseilles, France to pick up troops and take them to Hollandia and the Philippines. The *Sea Flier* stayed in the Pacific theater until March 1946 when she made another round trip to Le Havre and then one to Bremerhaven in June 1946. She was then released from troop carrying duty and returned to the USMC.

In 1948 the Luckenbach Steamship Company of New York purchased the *Sea Flier* from the USMC and changed her name to *Horace Luckenhach* and registered her in New York. This ship remained with Luckenbach from 1948 until 1961 when she was transferred to the Luckenbach Overseas Corporation under whom she operated from 1961 until 1967. In 1967 the *Horace Luckenhach* was transferred to Overseas Carrier Corporation of New York and her name was changed to *Overseas Horace*. In 1969 her name was again changed, this time to *Overseas Natalie*.

Disposal: The *Overseas Natalie* was sold to Taiwanese interests and scrapped at Kaohsiung in February 1971.

The Sea Flier *under construction on September 1, 1943. Looking from this view one gets a sense of just how complex a ship's hull is. Western Pipe & Steel Co., photo no. 765, author's collection.*

Sea Sturgeon

WPS Hull No. 124
USMC Hull No. 1547 Type C3-S-A2
Keel laid: August 28, 1943
Launched: December 7, 1943
Delivered: July 18, 1944.

The *Sea Sturgeon* was completed as a troop transport with a capacity of 2,114 troops and 231,686 cubic feet of cargo. She was operated by Waterman Steamship Company for the WSA.

The *Sea Sturgeon* departed San Francisco on her maiden voyage to Honolulu and South Pacific ports. She remained in Pacific service until February 1946 when she departed San Francisco for Liverpool and Le Havre. On return to New York the *Sea Sturgeon* underwent repairs and was placed in the USMC James River Reserve Fleet in early May 1946.

The Matson Navigation Company of San Francisco purchased this ship from the USMC November 27, 1947. She was renamed *Hawaiian Farmer* and carried twelve passengers and 10,446 tons of cargo between California and Hawaii. The service with Matson continued until April 12, 1971.

Disposal: In 1971 she was sold to L. Schnitzer American Ship Dismantlers, then was resold to Taiwanese interests and departed San Francisco under tow on June 9 for Kaohsiung, Taiwan to undergo dismantling.

This stern view of the Sea Sturgeon *was taken on November 2, 1943. The heavy framing which will support the stern is apparent. Western Pipe & Steel Co., photo no. 813, author's collection.*

Sea Runner

WPS Hull No. 125
USMC Hull No. 1548 Type C3-S-A2
Keel laid: September 29, 1943
Launched: January 4, 1944
Converted for use as a troopship during construction.
Delivered: June 29, 1944
Entered service: July 20, 1944.

The *Sea Runner* was operated by Grace Lines for the WSA. She carried 2,154 troops and 246,086 cubic feet of cargo.

The *Sea Runner* departed San Francisco July 23rd on her maiden voyage to Honolulu and Manus returning December 1944. She continued in service between the West Coast and points in the South and Western Pacific for the remainder of the War. After September 1945 there were trips to Japan and China as well as the Pacific islands. On her return to San Francisco in July 1946 the *Sea Runner* was placed in the USMC Reserve Fleet in Suisun Bay.

In 1949 the *Sea Runner* was purchased from the USMC by Luckenbach Steamship Company of New York. The ship was renamed *Robert Luckenbach* and registered in New York. This ownership continued from 1949 to 1961 when she was transferred to Overseas Carriers Corporation of New York and renamed *Overseas Rose*. From 1961 to 1971 the *Overseas Rose* was operated by Overseas Carrier Corporation.

Disposal: In 1971 *Overseas Rose* was sold to Taiwanese interests and scrapped at Kaohsiung in August 1971.

The Sea Runner *as a troop transport, tied to a pier in San Francisco on November 26, 1946. Author's collection.*

Sea Ray

WPS Hull No. 126
USMC Hull No. 1549 Type C3-S-A2
Keel laid: October 19, 1943
Launched: January 27, 1944

During construction this ship was converted for troop carrying. She had a capacity of 2,158 troops and 246,086 cubic feet of cargo. The *Sea Ray* was operated by American Hawaiian Steamship Company for the WSA.

The *Sea Ray* sailed on her first voyage in August 1944 to Guadalcanal and Espiritu Santo returning in September. From that time until March 1946 she continued in Pacific service. On return to San Francisco in March 1946 the *Sea Ray* was transferred and sailed to New York. On July 1, 1946 she was laid up in the USMC Reserve Fleet at Lee Hall, Virginia.

The *Sea Ray* was sold by the USMC to Matson Navigation Company of San Francisco on April 29, 1947 and was renamed *Hawaiian Rancher*. This ship carried twelve passengers and 10,446 tons of cargo in the trade between California and Hawaii. This service continued until June 15, 1971.

Disposal: In 1971 she was sold to Schnitzer Steel Products of Oakland, California for scrap. The *Hawaiian Rancher* was resold to Yung Tai Steel Corporation of Kaohsiung, Taiwan. She departed San Francisco under tow to Taiwan on July 7, 1972.

The ex-Sea Ray sailing for Matson Navigation Co. as the Hawaiian Rancher. She served Matson from 1947 to 1971. Matson Lines photo.

Sea Partridge

WPS Hull No. 127
USMC Hull No. 1550 Type C3-S-A2
Keel laid: November 27, 1943
Launched: February 29, 1944
This ship was completed as a troop transport.
Delivered: September 30, 1944.
Capacity:
    Cargo   185,446 cubic feet
    Troops   1,906

The *Sea Partridge* was operated by American President Lines for the WSA.

On October 2, 1944 this ship departed San Francisco on her first voyage to Puget Sound and Hawaii, returning in November. The *Sea Partridge* remained in Pacific service until December 1945. In addition to the usual ports in the Pacific islands she made trips between Seattle and the Alaska-Aleutian area. In February 1946 the *Sea Partridge* departed San Francisco for Liverpool and Le Havre returning to New York in March. In April 1946 this ship was released from troop transporting duties.

The USMC sold the *Sea Partridge* to Isthmian Steamship Company of New York in 1947. She was renamed *Steel Vendor* and registered in New York. Operation with Isthmian Steamship Company continued from 1947 to 1971.

On October 7, 1971 the *Steel Vendor* was lost when she was stranded 250 miles northeast of Manila.

The Steel Vendor *aground in the South China Sea. The helicoptor is from HMS* Eagle *and evacuated the crew. Shortly afterward the ship broke in two and flooded. U.S. Coast Guard.*

Sea Adder

WPS Hull No. 128
USMC Hull No. 1551 Type C3-S-A2
Keel laid: December 10, 1943
Launched: April 25, 1944
Name was changed to USS *Hansford* and subsequently to USS *Gladwin* (APA-106) on April 25, 1944.
Delivered: October 12, 1944

Commissioned in the U.S. Navy the same day she was completed, the *Gladwin* had a shakedown cruise to San Pedro and then departed November 25, 1944 for landing exercises in Hawaii. She spent the time until May 1946 in the Pacific theater shuttling troops. *Gladwin* took part in the Iwo Jima and Okinawa landings returning each time with casualties. On August 27 she departed Leyte for Tokyo Bay and the Japanese surrender. On return from Nagoya to Seattle with returning troops in April 1946 the *Gladwin* was sent to Norfolk, Virginia where she was decommissioned June 14, 1946 and returned to the USMC.

On May 20, 1947 the ex-*Gladwin* was sold to Isthmian Lines Inc. of New York and was renamed *Steel Apprentice*. Registered in New York she remained in the service of Isthmian Lines from 1949 to 1969.

Disposal: The *Steel Apprentice* was sold to Taiwanese interests in 1973 and in May 1973 was scrapped at Kaohsiung.

The Sea Adder *alongside the fitting out dock. Note the hull number on the bow. At this point she is yet to receive her cargo gear (although a crew can be seen rigging a topping lift on the forward starboard kingpost), armament and final painting. Western Pipe & Steel Co., photo no. 990, author's collection.*

Sea Wren

WPS Hull No. 129
USMC Hull No. 1552 Type C3-S-A2
Keel laid:　　　　January 7, 1944
Launched:　　　　May 31, 1944
During construction name was changed to
USS *Goodhue* (APA-107).
Delivered:　　　　November 11, 1944

Goodhue went into service December 1944 with a shakedown cruise in combination with practice amphibious landings. In March 1945 she participated in the Okinawa landings and sustained damage from a kamikaze with 27 killed and 117 injured. The rather superficial damage to the ship was repaired at Kerama Retto and *Goodhue* continued in service. In September she took troops to Sagami Bay, Japan, returning to Manila with liberated prisoners of war. Duty continued bringing servicemen back to the West Coast until February 1946 when the *Goodhue* was taken to Hampton Roads. Here she was decommissioned April 5, 1946 and returned to the USMC.

The former *Goodhue* was sold by USMC to Matson Navigation Company on April 15, 1947 and renamed *Hawaiian Citizen*. She carried twelve passengers and cargo between California and Hawaii. This ship underwent extensive alteration at Willamette Iron and Steel Company in Portland, Oregon between August 1959 and April 1960 becoming the first all-containerized freighter in service on the West Coast. The passenger accommodations were removed and net tonnage increased.

The *Hawaiian Citizen* was operated by Oceanic Steamship Company from March 8, 1971 until January 9, 1976 when she was returned to Matson service. This ship was laid up in San Francisco January 25, 1981.

Disposal: She was sold November 25, 1981 to Chi Shun Hua Steel Company of Kaohsiung, Taiwan for scrapping. The *Hawaiian Citizen* departed San Francisco on December 6, 1981 under tow for her final voyage to Kaohsiung for scrapping.

*The ex-*Sea Wren *became the* Hawaiian Citizen *for Matson Navigation Co. in 1947. In 1959 she was modified to carry containers, one of the first American ships to do so. Matson Navigation Co.*

Sea Hare

WPS Hull No. 130
USMC Hull No. 1553 Type C3-S-A2
Keel laid: January 31, 1944
Launched: June 29, 1944
Renamed USS *Goshen* (APA-108) during construction and completed as an attack transport.
Delivered: December 13, 1944
Commissioned: December 31, 1944

After a shakedown cruise along the California coast *Goshen* departed Long Beach for training in landing exercises in Hawaii February 4. 1945. She participated in Okinawa landings on April 17th and then shuttled troops around the Western Pacific. Was in service between the Philippines and Japan from September to December 1945. *Goshen* then returned a load of troops to San Diego in December and in January 1946 was sent to Lynnhaven Roads, Virginia arriving February 12th. This ship was decommissioned at Norfolk on April 20th and transferred to the WSA of the USMC May 2, 1946.

In 1947 the former *Goshen* was sold by the USMC to American Mail Lines, Ltd. of Seattle. The ship was renamed *Canada Mail* and registered in Portland, Oregon. In 1963 the name was changed to *California Mail*. She remained with American Mail Lines from 1947 until 1968 when she was sold to Waterman Steamship Company of New York. Her port of registry became New York and her name was changed to *LaFayette*. She was in service with Waterman Steamship Co. 1968 to 1973.

Disposal: In 1973 *LaFayette* was sold to Taiwanese interests and scrapped at Kaohsiung in August 1973.

Shown on a foggy day at an unidentified dock in the Hudson River, USS Goshen *(APA-108), ex-*Sea Hare, *awaits developments. Photograph Courtesy Peabody Essex Museum, neg. D-4908.*

Sea Sparrow

WPS Hull No. 131
USMC Hull No. 1554 Type C3-S-A2
Keel laid: March 3, 1944
Launched: August 10, 1944
Renamed USS *Grafton* (APA-109) during construction and completed as an attack transport.
Delivered: December 31, 1944
Commissioned: January 5, 1945

A shakedown cruise to San Pedro and amphibious training exercises in Southern California followed commissioning. On April 10 she departed and arrived Pearl Harbor April 16th. The *Grafton* spent the time until February 1946 shuttling troops, injured servicemen and prisoners of war about the Western Pacific. After returning to Seattle and San Francisco the ship was sent to Norfolk, Virginia where she was decommissioned May 16, 1946 and returned to the USMC on May 17th.

The USMC sold the former *Grafton* to American Mail Lines, Ltd. of New York in 1947. The name was changed to *Java Mail* and port of registry became Portland, Oregon.

She remained in the service of American Mail Lines from 1947 to 1969 when she was sold to Waterman Steamship Company of New York. The name was changed to *Carrier Dove* and she was in service with Waterman Steamship Company from 1969 to 1974.

Disposal: *Carrier Dove* was sold by Waterman Steamship Co. to Taiwanese interests in 1974 and arrived at Kaohsiung on May 28th for scrapping.

As USS Grafton *(APA-109), upper, the ex-*Sea Sparrow *is nested with other ship at the James River Reserve Fleet. Below she is shown under construction on July 4, 1944. Note how many workers are at different locations on the ship. Upper, Photo Courtesy Peabody Essex Museum, neg. D-4909; lower, Western Pipe & Steel Co., photo no. 980, author's collection.*

Sea Cardinal

WPS Hull No. 132
USMC Hull No. 1555 Type C3-S-A2
Keel laid:　　　April 28, 1944
　　Launching way number 2
Launched:　　　September 7, 1944
Delivered:　　　January 29,1945.

Operated by Luckenbach Steamship Company for the WSA in 1945-46.

The *Sea Cardinal* was sold by the USMC to Isthmian Steamship Company of New York in 1946. She was renamed *Steel Architect* and registered in New York. This ship remained with Isthmian Steamship Co. from 1946 to 1971.

Disposal: In 1971 *Steel Architect* was sold to Taiwanese interests and scrapped at Kaosiung in August of that year.

The Sea Cardinal *under construction. Note the carefully laid out chalk marks on the deck indicating location of part of deck house; women shipyard workers in foreground. Western Pipe & Steel Co., photo no. 982, author's collection.*

Sea Finch

WPS Hull No. 133
USMC Hull No. 1556 Type C3-S-A2
Keel laid: June 3, 1944
       Launching way number 4
Launched: October 5, 1944
Name was changed to *Sea Shark* during construction.
Delivered: February 17, 1945.

The *Sea Shark* was operated by Weyerhaeuser Steamship Company for the WSA during 1945-46.

The *Sea Shark* was sold by the USMC to Isthmian Steamship Company of New York in 1946. She was renamed *Steel Maker* and remained in service with Isthmian Steamship Co. from 1946 to 1973 with New York as her port of registry.

Disposal: In 1973 the *Steel Maker* was sold to Taiwanese interests and was scrapped at Kaohsiung July 1973.

This bow view of the Sea Finch *taken on July 4, 1944 clearly shows the closely spaced heavy framing necessary to strengthen the forward part of a ship's hull. Western Pipe & Steel Co., photo no. 983, author's collection.*

Sea Starling (I)

WPS Hull No. 134
USMC Hull No. 1557 Type C3-S-A2
Keel laid: July 3, 1944
    Launching way number 1.
Launched: November 8, 1944
Name was changed to *Sea Blenny* during construction.
Delivered: March 9, 1945.

The *Sea Blenny* was operated by Pope & Talbot Inc. for the WSA for the 1945-46 period. As indicated by the photo on the opposite page the *Sea Blenny* was apparently operated by APL for a short period before the sale to Matson Navigation Company.

The USMC sold the *Sea Blenny* to Matson Navigation Company on March 18, 1947 and her name was changed to *Hawaiian Builder*. She carried 10,446 tons of cargo between California and Hawaii. Matson sold *Hawaiian Builder* back to the USMC June 23, 1970 and this ship continued to be operated by Matson under a "use" agreement until December 1970. In December 1970 the ship was put in the Marad Reserve Fleet in Suisun Bay. She was sold to Inter-Ocean Grain Storage Ltd. on June 26, 1973.

Disposal: In September 1973 the ship was sold to Taiwanese interests and scrapped at Kaohsiung in November.

Shown in an early stage of construction the Sea Blenny*'s stern plating begins to take form. Western Pipe & Steel Co., photo no. 986, author's collection.*

The Sea Blenny *displays American President Lines markings as a freighter. Note the glassy sea reflecting brightly off the hull at no. 1 hold. Photo Courtesy Peabody Essex Museum, neg. D-3551.*

Sea Thrush

WPS Hull No.135
USMC Hull No. 2188 Type C3-S-A4
Tonnage:
 Gross: 8,000
 Deadweight: 11,000
Length overall: 492 feet
Beam: 69 feet 6 inches
Maximum draft: 26 feet 6 inches
Machinery:
 2 geared turbines
Speed: 16.5 knots
Keel laid   August 14, 1944
 Launching way number 3
Launched: December 7, 1944
Delivered: July 20, 1945

After delivery to the USMC *Sea Thrush* was sold to the American President Lines and was renamed *President Taft*, being the second vessel so named by the APL. She carried cargo and twelve passengers. On February 28, 1968 this ship was renamed *President Harding* and was the fourth vessel so named by APL.

On January 26, 1973 this ship was sold to Bonito Maritime Corporation and renamed *Harding*.

Disposal: In 1973 the *Harding* was sold to Taiwanese interests and taken there for scrapping.

After being sold to American President Lines, the Sea Thrush *became the* President Taft *(II) then the* President Harding *(IV). Despite the fact that the photo on the opposite page was taken twenty-three years before, the ship appears to be in remarkably good condition, above. Both photos, author's collection.*

PRESIDENT TAFT

Sea Beaver

WPS Hull No. 136
USMC Hull No. 2189 Type C3-S-A4
Tonnage:

Gross:	8,000
Deadweight:	11,000

Length overall: 492 feet
Beam: 69 feet 6 inches
Maximum draft: 26 feet 6 inches
Machinery:
  2 geared turbines
Speed: 16.5 knots
Keel laid: September 11, 1944
  Launching way number 2
Launched: January 8, 1945

The *Sea Beaver* was delivered to the USMC August 8, 1945 and had no war service.

The *Sea Beaver* was sold to American President Lines and renamed *President Grant*, being the third vessel so named by APL. She carried cargo and twelve passengers. On September 26, 1967 this ship was renamed *President Hoover*.

On October 27, 1972 the *President Hoover* was sold to Excelsior Marine Corporation and renamed *Hoover*.

Disposal: In 1973 the *Hoover* was sold to Taiwanese interests and taken there for scrapping.

The interior photo of the Sea Beaver, *lower right, shows the working tunnel for cargo oil, located directly above the double bottom and running from the forward engine room bulkhead to the forward end of No. 2 hold. Upper photo shows the vessel as a well-used* President Hoover *in San Francisco on October 24, 1967. Both from author's collection.*

Sea Jumper

WPS Hull No. 137
USMC Hull No. 2190 Type C3-S-A4
Tonnage:

Gross:	8,000
Deadweight:	11,000

Length overall: 492 feet
Beam: 69 feet 6 inches
Maximum draft: 26 feet 6 inches
Machinery:
   2 geared turbines
Speed: 16.5 knots
Keel laid: October 9, 1944
   Launching way number 4
Launched: February 7, 1945
Delivered: September 18, 1945

The *Sea Jumper* was sold by USMC to American President Lines and the ship's name was changed to *President Pierce*, being the second ship so named by APL. The *President Pierce* carried cargo and twelve passengers.

On December 14, 1972 this ship was sold to Amber Jack Maritime Corporation and renamed *Pierce*.

Disposal: In 1973 the *Pierce* was sold to Taiwanese interests and scrapped in Kaohsiung.

Taken in San Francisco on November 29, 1946, this photo shows the President Pierce (II), *ex-*Sea Jumper. *At this point in her career she lacks the dents, rust and other indications of a well-used ship. Author's collection.*

Sea Starling (II)

WPS Hull No. 138
USMC Hull No. 2191 Type C3-S-A4
Tonnage:
    Gross: 8,000
    Deadweight: 11,000
Length overall: 492 feet
Beam: 69 feet 6 inches
Maximum draft: 26 feet 6 inches
Machinery:
    2 geared turbines
Speed: 16.5 knots
Keel laid: November 14,1944
    Launching way number 1
Launched: March 6, 1945
Delivered: June 12, 1946

Sold shortly after delivery to American President Lines by USMC. The *Sea Starling* was renamed *President Madison* and was the third ship so named by APL. She carried twelve passengers and cargo. On August 7, 1972 she was sold to Vintage Steamship Steamship Company and renamed *Madison*.

Disposal: The *Madison* was sold to Taiwanese interests in 1972 and taken there for scrapping.

Right, the Sea Starling *before launching. Note the hull number still painted forward of the vessel's name. Below, as the* President Pierce *(III) she steams on a calm sea with little or no cargo (note boot-topping). Right, author's collection, below American President Lines Archives.*

Sea Phoebe

WPS Hull No. 175
USMC Hull No. 2192 Type C3-S-A4
Tonnage:
    Gross: 8,000
    Deadweight: 11,000
Length overall: 492 feet
Beam: 69 feet 6 inches
Maximum draft: 26 feet 6 inches
Machinery:
    2 geared turbines
Speed: 16.5 knots
Keel laid: December 11, 1944
    Launching way number 3
Launched: April 12, 1945
Delivered: July 19, 1946

Shortly after delivery to USMC, the vessel was sold to American President Lines. She was renamed *President McKinley*, being the second ship so named by APL. She carried twelve passengers and cargo. On April 11, 1968 this ship had its name changed to *President Johnson*.

On December 6, 1969 the *President Johnson* was sold to Pinedale Shipping Company and renamed *Pinedale*.

Disposal: In 1970 the *Pinedale* was sold to Taiwanese interests and taken there for scrapping.

The launching of the Sea Phoebe, above right, drew the rapt attention of a large crowd on April 12, 1945. Later, as the President McKinley (II), below, she served APL for more than twenty years. Note the light colored hull below. The company went to black hulls shortly after this 1947 photo was taken. Both author's collection.

Sea Oriole

WPS Hull No. 176.
USMC Hull No. 2193 Type C3-S-A4
Tonnage:
    Gross: 8,000
    Deadweight: 11,000
Length overall: 492 feet
Beam: 69 feet 6 inches
Maximum draft: 26 feet 6 inches
Machinery:
    2 geared turbines
Speed: 16.5 knots
Keel laid: January 11, 1945
    Launching way number 2
Launched: May 17, 1945
Delivered: August 22, 1946

Shortly after delivery to the USMC the vessel was sold to American President Lines. The *Sea Oriole* was renamed *President Jefferson* and was the second vessel so named by APL. The *President Jefferson* (II) carried cargo and twelve passengers.

On February 20, 1970 this ship was sold to Ferndale Shipping Company and was renamed *Ferndale*.

Disposal: In 1970 she was sold to Taiwanese interests and taken to Kaohsiung for scrapping.

December 10, 1946 and the President Jefferson *(II) backs away from a San Francisco pier with the Bay Bridge in the background. Note the twin-stack steam tug in the left foreground. Author's collection.*

The President Jefferson *(II) steaming across San Francisco Bay showing a light "economy" haze coming from her funnel and a clean profile. American President Lines Archives.*

Sovereign of the Seas

WPS Hull No. 244
USMC Hull No. 1157 Type C2-S-B1
Tonnage:
    Gross: 6,230
    Deadweight: 9,150
Length overall: 429 feet
Beam: 62 feet
Machinery:
    geared turbine
    6,000 SHP
Speed: 16.5 knots

The hull of this ship was constructed and launched at Moore Drydock Company of Oakland, California. The hull was then taken over and completed by WPS Company. This ship was delivered to the USMC on February 29, 1944. The *Sovereign of the Seas* was operated by Isthmian Steamship Company for the WSA in the 1944-46 period with the homeport being San Francisco.

In 1947 the ship was sold to Agwilines Inc. of New York and New York became her port of registry.

In 1948 *Sovereign of the Seas* was sold to New York & Cuba Steamship Company of New York and renamed *Agwidale*. The ship was owned by this company from 1948 to 1954. Her name was again changed to *Oriente* in 1950.

In 1954 *Oriente* was sold to A.H. Bull Steamship Company of New York. She was renamed *Short Hills* briefly and then *Jean*. The *Jean* was owned by A.H. Bull Co. from 1954 to 1964.

In 1964 the *Jean* was sold to Oceanic Ore Carriers Inc. of New York and the name was changed to *Oceanic Tide* and she was registered in Wilmington, Delaware. Service with this company continued from 1964 to 1968 when *Oceanic Tide* was sold to Resolute Marine Association. This company operated the ship from 1968 to 1969.

Disposal: In 1969 *Oceanic Tide* was sold to Taiwanese interests and scrapped at Kaohsiung March 1969.

White Swallow

WPS Hull No. 245
USMC Hull No. 1158 Type C2-S-B1
Tonnage:
    Gross: 6,230
    Deadweight: 9,150
Length overall: 429 feet
Beam: 62 feet
Machinery:
    geared turbine
    6,000 SHP
Speed: 16.5 knots

The hull was constructed and launched by Moore Drydock Company of Oakland California. The hull was then taken over and completed by WPS Company. The *White Swallow* was delivered in April 1944 to the USMC and was operated by Moore-McCormack Line for the WSA during the 1944-46 period with San Francisco as port of registry.

The USMC sold *White Swallow* to Moore-McCormack Lines Inc. of New York in 1947. She was renamed *Mormacowl* and remained in service with Moore-McCormack Lines from 1947 to 1965 with New York as port of registry.

In 1965 *Mormacowl* was purchased by Sperling Steamship & Trading Corporation of New York. The ship was renamed *Old Westbury* and she was operated by this company from 1965 to 1967.

In 1967 the *Old Westbury* was sold to Westbury Steamship Corporation which operated the vessel from 1967 to 1969.

Disposal: In 1969 *Old Westbury* was sold to Taiwanese interests and taken to Kaohsiung for scrapping in July 1969.

The White Swallow *under way as the* Mormacowl *during her service with Moore-McCormack Lines.* Photo: V.H. Young/L.A. Sawyer.

NOTE:
WPS Hull numbers 23-57, 66-76, 96-120, 139-174, and 177-243 are unaccounted for in the WPS records for their South San Francisco yard. As stated earlier it is presumed that hull numbers 23-57 represent barges built in the 1920-38 time period and numbers 66-76 were a group of small barges built about 1939. Numbers 96-120 were destroyer escorts built by the San Pedro yard, 139 was a cancelled C-3, 140-174 were Coast Guard cutters constructed at San Pedro and numbers 177-243 were LSMs which were cancelled.

At the time of writing this history all the ships have been scrapped. The only exceptions are ships lost by accidents or sunk in action during World War II.

Appendices

Bibliography

Index

APPENDIX I
WORLD WAR I VESSELS

Ship Name	WPS Hull No.	Gross Tonnage	Delivered
Isanti	1	5713	October 6, 1918
Nantahala	2	5714	October 31, 1918
West Avenal	3	5692	February 1, 1919
Oskaloosa	4	5623	December 20,1918
West Vaca	5	5548	March 27, 1919
West Ashawa	6	5609	May 21, 1919
West Alcoz	7	5606	June 24, 1919
West Aleta	8	5605	August, 1919
West Cactus	9	5643	August, 1919
West Caddoa	10	5641	September 10, 1919

Ship Name	WPS Hull No.	Gross Tonnage	Delivered
West Kader	11	5570	December 31, 1919
West Cadron	12	5724	March 3,1920
West Cahokia	13	5645	April 16, 1920
West Calera	14	5644	January 14, 1920
West Camak	15	5647	__________, 1920
West Camargo	16	5881	July 21, 1920
West Canon	17	5645	September, 1920
West Carmona	18	5645	October, 1920

Note: Four ships were cancelled at the end of World War I

Appendix II
Vessels built for World War II

Ship Name	WPS Hull No.	USMC Hull No.	Use	Sponsor
American Manufacturer	57	94	C-1 cargo	Mrs. Kenneth Dawson
American Leader	58	95	C-1 cargo	Mrs. John H. Tolan
American Builder	59	96	C-1 cargo	Miss Aileen Tallerday
American Press	60	97	C-1 cargo	Miss Camilla Chandler
American Packer	61	98	C-1 cargo	Mrs. Angelo J. Rossi
HMS *Attacker*	62	171	CVE	Mrs. W.A. Ross
USS *Cascade*	63	172	Destroyer Tender	Mrs. C.W. Crosse
USS *Chandeleur*	64	173	Seaplane Tender	Mrs. Frank McCrary
HMS *Stalker*	65	174	CVE	Mrs. W.H. Shea

Ship Name	WPS Hull No.	USMC Hull No.	Use	Sponsor
HMS *Fencer*	77	197	CVE	Mrs. Powers Symington
HMS *Striker*	78	198	CVE	Mrs. Don Bosh
Sea Pike	79	267	USMC transport	Mrs. John L. Patton
Sea Bass	80	268	USMC transport	Mrs. Clarence A. Reiter
USS *Bolivar*	81	269	Navy transport	Mrs. Robert W. Ethen
USS *Callaway*	82	270	Navy transport	Mrs. W.P. Manuell
USS *Cambria*	83	271	Navy transport	Mrs. W.H. Griffin
Sea Snipe	84	272	USMC transport	Mrs. Frank F. Kane
USS *Chilton*	85	273	Navy transport	Mrs. W.A. Riley, Jr.
USS *Clay*	86	274	Navy transport	Mrs. Earl Warren
USS *Bayfield*	87	275	Navy transport	Mrs. J.E. Schmeltzer
USS *Cavalier*	88	276	Navy transport	Mrs. M.J. Jackson
Sea Barb	89	277	Army Transport	Mrs. David Currier
Sea Cat	90	278	Army Transport	Mrs. C.W. Flesher
Sea Devil	91	279	USMC Transport	Mrs. Maxwell McNutt
Sea Flasher	92	280	USMC Transport	Mrs. Christian Andersen

Ship Name	WPS Hull No.	USMC Hull No.	Use	Sponsor
USS *Alpine*	93	281	Navy transport	Mrs. Helen March
USS *Barnstable*	94	282	Navy transport	Miss Jean Watts
Sea Corporal	95	283	Army transport	Mrs. Fern Larsen
USS *Cecil*	121	1544	Navy transport	Mrs. J.L. Bloom
Sea Fiddler	122	1545	Army transport	Mrs. Jean Mehrrnann
Sea Flier	123	1546	Army transport	Miss Clair Nooning
Sea Sturgeon	124	1547	Army transport	Mrs. Leo Carrillo
Sea Runner	125	1548	Army transport	Mrs. John A. McKeown
Sea Ray	126	1549	Army transport	Mrs. Mary Mangan
Sea Partridge	127	1550	Army transport	Mrs. John R. Reilly
USS *Gladwin*	128	1551	Navy transport	Mrs. E.G. Cahill
USS *Goodhue*	129	1552	Navy transport	Mrs. Charles H. Purcell
USS *Goshen*	130	1553	Navy transport	Mrs. James B. Black
USS *Grafton*	131	1554	Navy transport	Mrs. S. Belither
Sea Cardinal	132	1555	C-3 cargo	Mrs. L.L. Cohen
Sea Shark	133	1556	C-3 cargo	Mrs. K.U. Plummer

Ship Name	WPS Hull No.	USMC Hull No.	Use	Sponsor
Sea Blenny	134	1557	C-3 cargo	Mrs. Nelson Eckart
President Taft	135	2188	C-3 cargo-passenger	Mrs. L.D. Jurs
President Grant	136	2189	C-3 cargo-passenger	Mrs. Barbara Quarg
President Pierce	137	2190	C-3 cargo-passenger	Mrs. Earl C. Elliott
President Madison	138	2191	C-3 cargo-passenger	Mrs. T.S. Petersen
President McKinley	175	2192	C-3 cargo-passenger	Mrs. J.W. Fowler
President Jefferson	176	2193	C-3 cargo-passenger	Mrs. A.E. Wishon
Sovereign of the Seas	244	1157	C-2 cargo	---------
White Swallow	245	1158	C-2 cargo	---------

APPENDIX III
WORLD WAR II VESSELS
SIGNIFICANT DATES

Ship's Name	Keel Laid	Launched	Delivered
American Manufacturer	2/5/40	8/8/40	4/11/41
American Leader	2/19/40	10/8/40	6/12/41
American Builder	8/15/40	10/17/40	7/25/41
American Press	11/11/40	3/11/41	9/3/41
American Packer	12/23/40	5/20/41	10/14/41
HMS *Attacker*	4/7/41	9/27/41	9/30/42
USS *Cascade*	7/17/41	6/7/42	9/11/42

Ship's Name	Keel Laid	Launched	Delivered
USS *Chandeleur*	5/29/41	11/29/41	11/19/42
HMS *Stalker*	10/6/41	3/5/42	12/21/42
HMS *Fencer*	9/5/41	4/4/42	2/27/43
HMS *Striker*	12/15/41	5/7/42	4/28/43
Sea Pike	3/9/42	6/28/42	2/13/43
Sea Bass	4/10/42	8/2/42	3/31/43
USS *Bolivar*	5/13/42	9/7/42	3/15/43
USS *Callaway*	6/10/42	10/10/42	4/24/43
USS *Cambria*	7/1/42	11/10/42	5/4/43
Sea Snipe	8/8/42	12/7/42	5/29/43
USS *Chilton*	9/10/42	12/29/42	5/29/43
USS *Clay*	10/14/42	1/23/43	6/29/43
USS *Bayfield*	11/14/42	2/15/43	6/30/43
USS *Cavalier*	12/10/42	3/15/43	7/19/43
Sea Barb	12/31/42	4/8/43	8/6/43
Sea Cat	1/26/43	5/4/43	8/25/43
Sea Devil	2/18/43	5/23/43	11/30/43

Ship's Name	Keel Laid	Launched	Delivered
Sea Flasher	3/18/43	6/22/43	12/24/43
USS *Alpine*	4/12/43	7/10/43	9/30/43
USS *Barnstable*	5/6/43	8/5/43	10/30/43
Sea Corporal	5/25/43	8/26/43	1/31/44
USS *Cecil*	6/24/43	9/27/43	2/26/44
Sea Fiddler	7/13/43	10/16/43	5/18/44
Sea Flier	8/7/43	11/16/43	5/27/44
Sea Sturgeon	8/28/43	12/7/43	7/18/44
Sea Runner	9/29/43	1/4/44	6/29/44
Sea Ray	10/19/43	1/27/44	7/29/44
Sea Partridge	11/27/43	2/29/44	9/30/44
USS *Gladwin*	12/10/43	4/25/44	10/12/44
USS *Goodhue*	1/7/44	5/31/44	11/11/44
USS *Goshen*	1/31/44	6/29/44	12/13/44
USS *Grafton*	3/3/44	8/10/44	12/31/44
Sea Cardinal	4/28/44	9/7/44	1/29/45
Sea Shark	6/3/44	10/5/44	2/17/45

Ship's Name	Keel Laid	Launched	Delivered
Sea Blenny	7/3/44	11/8/44	3/9/45
President Taft	8/14/44	12/7/44	7/20/45
President Grant	9/11/44	1/8/45	8/8/45
President Pierce	10/9/44	2/7/45	9/18/45
President Madison	11/14/44	3/6/45	6/12/46
President McKinley	12/11/44	4/12/45	7/29/46
President Jefferson	1/11/45	5/17/45	8/22/46

BIBLIOGRAPHY

Blum, Joseph A. "South San Francisco: The Making Of An Industrial City." *California History*, Volume 63, Number 2. San Francisco: California Historical Society, Spring 1984.

Charles, Roland W. *Troopships of World War II*. Washington, D.C.: The Army Transport Association, 1947.

Kauffman, Linda. *South San Francisco, A History*. South San Francisco, California: City of South San Francisco, 1976.

Lloyd's Register of Shipping. London, U.K.: Lloyd's Register of Shipping, 1920 to 1975 annual editions.

Maslin, Marshall, Ed. *Western Shipbuilders in World War II, A Detailed Review of Wartime Activities of Leading Maritime and Navy Contractors*. Oakland, California: Shipbuilding Review Publishing Association, nd. (circa 1946).

Merchant Vessels of the United States. Washington, D.C.: Department of Commerce, Bureau of Navigation, 1925 to 1975 annual editions.

Moulin, Gabriel. *Schaw Batcher Shipyard*. A bound group of panoramic photographs depicting the construction of the shipyard, 1918.

Poolman, Kenneth. *Escort Carrier 1941-1945, An Account of British Escort Carriers in Trade Protection*. London, U.K.: Ian Allan, 1972.

Poolman, Kenneth. *Escort Carriers of World War Two*. London, U.K.: Fotofax Arms and Armour Press, 1989.

Record of American Bureau of Shipping. New York, N.Y.; American Bureau of Shipping, 1920 to 1975 annual editions.

Roy, Rose. *Ships in San Mateo County, Their Cradles - Their Ports of Call*. Student monograph #323 in the archives of the San Mateo County Historical Museum, 1941.

Sawyer, L.A. and W.H. Mitchell. *From America to United States*. Vol. 1. Kendal: World Ship Society, 1979.

Sawyer, L.A. and W.H. Mitchell. *From America to United States*. Vol. 2. Kendal: World Ship Society, 1981.

Sawyer, L.A. and W.H. Mitchell. *From America to United States*. Vol. 3. Kendal: World Ship Society, 1984.

Schmidt, Earl F. "History of the Meat Industry In San Mateo County." *La Peninsula*. Journal of the San Mateo Country Historical Association. Volume 18 Number 2. San Mateo: San Mateo County Historical Society, Spring 1976.

Stanger, Frank M. "Ships and Ship Building." *La Peninsula*. Journal of the San Mateo County Historical Association, Vol. 2, No. 1. San Mateo: San Mateo County Historical Society, February 1943.

Stanger, Frank M. *South From San Francisco, San Mateo County, California, Its History and Heritage*. San Mateo, California: San Mateo County Historical Association, 1963.

Western Pipe & Steel Company of California Fabricators & Erectors of Steel Products, An Illustrated Booklet of Plant & Products. San Francisco, California: Western Pipe & Steel Company of California, 1944.

Western Pipe & Steel News. A twice monthly plant newspaper. San Francisco, California: First issue appeared September 1, 1942. Last issue known to author is October 1, 1945. It may have continued until the sale of the yard in January 1946.

Index

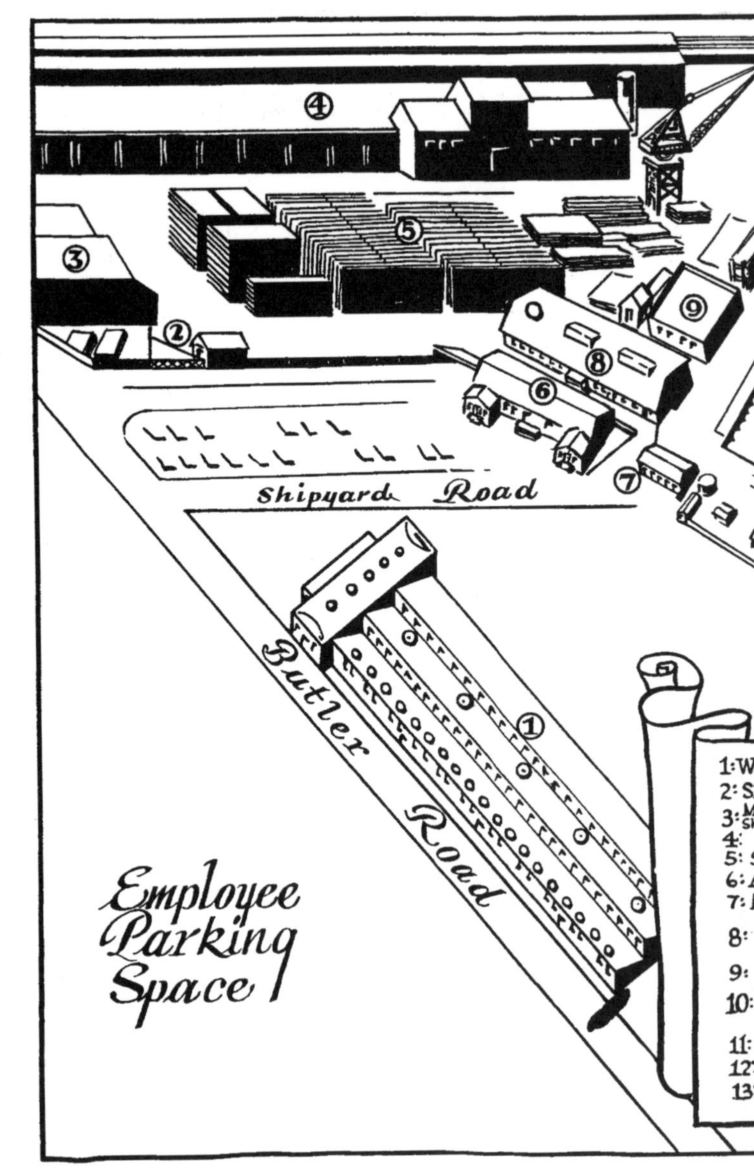

Shipyard Road
Butler Road
Employee Parking Space
1: Warehouse
2: Shop Gate
3: Machine
Sheet Meta
4: Shops
5: Steel Yar
6: Adm. Bl
7: Main Gat
8: Cafeter
Class Ro
First Ai
9: Time Kee
10: Yard & Wa
Welding
Offices,
11: Way #4
12: Way #3
13: Way #1